Using Operation-Based Multi-Queue SBC Process Algebra as a Metamodel for UML

-- Toward a Unified View of the System --

William S. Chao • Shuh-Ping Sun

Structure-Behavior Coalescence

Software Architecture = Software Structure + Software Behavior

CONTENTS

CONTENTS 5

PREFACE 11

ABOUT THE AUTHORS 13

PART I: BASIC CONCEPTS 15

Chapter 1: Why Does UML Need a Metamodel? 17

1-1 Unified Modeling Language 17

1-2 Inconsistency Problems of UML 18

1-3 Metamodel of UML 19

1-4 Deficiency of Current UML Metamodels 20

1-5 SBC Approach for UML Metamodel 21

Chapter 2: Introduction to Process Algebra 23

2-1 What is Process Algebra? 23

2-2 Well-known Examples of Process Algebra 23

2-3 Generalized SBC Process Algebra 23

2-4 Specialized SBC Process Algebras 24

Chapter 3: Mathematics of Process Algebra 25

3-1 Sequentialization of Interactions 25

3-2 Summation of Processes 25

3-3 Parallel Composition of Processes 25

3-4 Recursive Definition of a Process 25

3-5 Replication of a Process 26

3-6 Conditional Definition of a Process 26

3-7 Null Process 26

PART II: OPERATION-BASED MULTI-QUEUE SBC PROCESS ALGEBRA 29

Chapter 4: Operation-Based Value-Passing Interactions 31

4-1 Operations 31

4-2 Operation-Based Value-Passed Interactions 32

4-3 Formal Description of an Operation-Based Interaction 37

Chapter 5: The Structure-Behavior Coalescence Approach 39

5-1 Structure-Behavior Coalescence Means to Integrate the Software Structure and Software Behavior 39

5-2 Interactions among Objects and the External Environment to Draw Forth the Software Behavior 40
5-3 Core Theme of Structure-Behavior Coalescence..................................42
Chapter 6: Language Constructs of Operation-Based Multi-Queue SBC Process Algebra 43
6-1 Entity Set and Entity Name 43
6-2 Backus-Naur Form of Operation-Based Multi-Queue SBC Processes .44
6-2-1 Parallel Composition of One or More Recursive Interaction Flow Diagrams Defines the Operation-Based Multi-Queue SBC Process of a System 45
6-2-2 A Recursive Interaction Flow Diagram is the Recursion of an Interaction Flow Diagram 45
6-2-3 An Interaction Flow Diagram is a Type_1 Interaction Followed by Zero or More Type_1_Or_Type_2 Interactions.............................46
6-2-4 Type_1_Or_2 Interaction is either Type_1 or Type_2..............46
Chapter 7: Transitional Semantics of Operation-Based Multi-Queue SBC Process Algebra 49
7-1 Transitional Semantics...49
7-2 Rule of Prefix...50
7-3 Rule of Summation 51
7-4 Rule of Recursion 51
7-5 Rule of Parallel Composition...52
7-6 Rule of Constants...53
PART III: MAPPING FROM O-M-SBC-PA TO UML.....................55
Chapter 8: O-M-SBC-PA Transition Graph.......................................57
8-1 Definition of a System in O-M-SBC-PA...57
8-2 Transition Graphs in O-M-SBC-PA 57
8-3 TG Relations (TGR) in O-M-SBC-PA 58
8-4 Orthogonal Composite State in the Transition Graph 58
8-5 Transition Graph of FixIFD...61
8-6 TG Relation of FixIFD 61
8-7 Transition Graph of a System..62
8-8 TG Relation of a System...63
Chapter 9: Projecting a Use Case Diagram from the O-M-SBC-PA Transition Graph 65
9-1 UML Use Case Diagrams 65
9-2 UCD Relation (UCDR) of a System...66
9-3 Algorithm of Projecting a Use Case Diagram from O-M-SBC-PA 66

Chapter 10: Projecting a State Diagram from the O-M-SBC-PA Transition Graph 69

10-1 UML State Diagrams 69

10-2 StD Relation (StDR) of a System 70

10-3 Algorithm of Projecting a State Diagram from O-M-SBC-PA 70

Chapter 11: Projecting an Activity Diagram from the O-M-SBC-PA Transition Graph 73

11-1 UML Activity Diagrams 73

11-2 AD Relation (ADR) of a System 74

11-3 Algorithm of Projecting an Activity Diagram from O-M-SBC-PA 75

Chapter 12: Projecting a Sequence Diagram from the O-M-SBC-PA Transition Graph 79

12-1 UML Sequence Diagrams 79

12-2 SqD Relation (SqDR) of a System 81

12-3 Algorithm of Projecting a Sequence Diagram from O-M-SBC-PA 83

Chapter 13: Projecting a Communication Diagram from the O-M-SBC-PA Transition Graph 85

13-1 UML Communication Diagrams 85

13-2 ComD Relation (ComDR) of a System 87

13-3 Algorithm of Projecting a Communication Diagram from O-M-SBC-PA 89

Chapter 14: Projecting a Class Diagram from the O-M-SBC-PA Transition Graph 91

14-1 UML Class Diagrams 91

14-2 ClsD Relation (ClsDR) of a System 91

14-3 Algorithm of Projecting Class Diagrams from O-M-SBC-PA 92

Chapter 15: Projecting an Object Diagram from the O-M-SBC-PA Transition Graph 95

15-1 UML Object Diagrams 95

15-2 OD Relation (ODR) of a System 96

15-3 Algorithm of Projecting an Object Diagram from O-M-SBC-PA....96

PART IV: CASE STUDY 99

Chapter 16: Online Shopping Systems 101

16-1 O-M-SBC-PA Process of the Online Shopping System 101

16-2 O-M-SBC-PA Transition Graph of the Online Shopping System....101

16-3 TG Relation of the Online Shopping System 103

Chapter 17: Projecting a Use Case Diagram from the O-M-SBC-PA Transition Graph of the Online Shopping System 107

17-1 Projecting the UCD Relation from the Transition Graph of the Online Shopping System....107
17-2 Achieving the Use Case Diagram from the UCD Relation of the Online Shopping System....108
Chapter 18: Projecting a State Diagram from the O-M-SBC-PA Transition Graph of the Online Shopping System....111
18-1 Projecting the StD Relation from the Transition Graph of the Online Shopping System....111
18-2 Achieving the State Diagram from the StD Relation of the Online Shopping System....115
Chapter 19: Projecting an Activity Diagram from the O-M-SBC-PA Transition Graph of the Online Shopping System....117
19-1 Projecting the AD Relation from the Transition Graph of the Online Shopping System....117
19-2 Achieving the Activity Diagram from the AD Relation of the Online Shopping System....121
Chapter 20: Projecting a Sequence Diagram from the O-M-SBC-PA Transition Graph of the Online Shopping System....123
20-1 Projecting the SqD Relation from the Transition Graph of the Online Shopping System....123
20-2 Achieving the Sequence Diagram from the SqD Relation of the Online Shopping System....127
Chapter 21: Projecting a Communication Diagram from the O-M-SBC-PA Transition Graph of the Online Shopping System....131
21-1 Projecting the ComD Relation from the Transition Graph of the Online Shopping System....131
21-2 Achieving the Communication Diagram from the ComD Relation of the Online Shopping System....135
Chapter 22: Projecting a Class Diagram from the O-M-SBC-PA Transition Graph of the Online Shopping System....139
22-1 Projecting the ClsD Relation from the Transition Graph of the Online Shopping System....139
22-2 Achieving the Class Diagram from the ClsD Relation of the Online Shopping System....141
Chapter 23: Projecting an Object Diagram from the O-M-SBC-PA Transition Graph of the Online Shopping System....143
23-1 Projecting the OD Relation from the Transition Graph of the Online Shopping System....143

23-2 Achieving the Object Diagram from the OD Relation of the Online Shopping System....................144

BIBLIOGRAPHY 147

INDEX 153

PREFACE

Unified Modeling Language (UML) is a general modeling language for model-driven (software) engineering (MDE) applications. The UML specification defines a set of language concepts that is used to model the (static) structure and (dynamic) behavior of a system. The UML concepts include (1) an abstract syntax that defines the language concepts and is described by a metamodel, and (2) a concrete syntax, or notation, that defines how the language concepts are represented and is described by a user model.

Since UML is a multi-diagram approach, there are always some inconsistencies between different diagrams in the user model. To ensure and check the consistency, a metamodel that defines the abstract syntax of a modeling language needs to provide a unified semantic framework for defining consistency rules to impose constraints on the structure (i.e., objects) or behavior (i.e., activities) constructs. It is hoped that through this unified semantic framework, each diagram in the user model can be projected as a view of the metamodel.

Unfortunately, most current UML metamodels do not have the ability to project each diagrams in the user model as a view of the metamodel. In this book, we develop the Operation-Based Multi-Queue Structure-Behavior Coalescence Process Algebra (O-M-SBC-PA) as the metamodel of UML. In O-M-SBC-PA, each diagram in the user model will be projected as a view of the metamodel. Therefore, we claim that O-M-SBC-PA genuinely provides a unified semantic framework to ensure model consistency for UML.

ABOUT THE AUTHORS

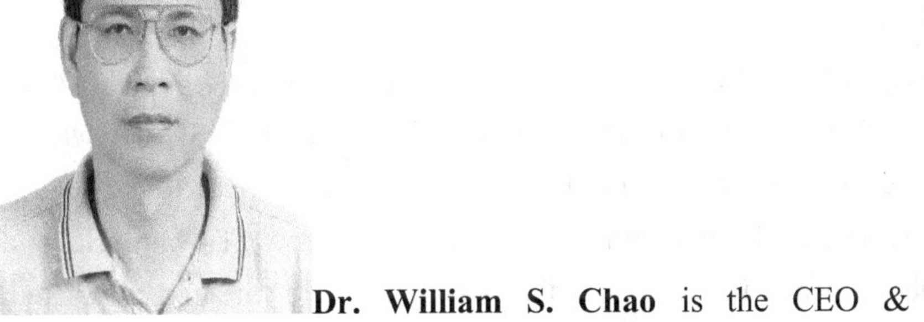

Dr. William S. Chao is the CEO & founder of SBC Architecture International®. SBC (Structure-Behavior Coalescence) architecture is a systems architecture which demands the integration of systems structure and systems behavior of a system. SBC architecture applies to hardware architecture, software architecture, enterprise architecture, knowledge architecture and thinking architecture. The core theme of SBC architecture is: Architecture = Structure + Behavior.

William S. Chao received his bachelor degree (1976) in telecommunication engineering and master degree (1981) in information engineering, both from the National Chiao-Tung University, Taiwan. From 1976 till 1983, he worked as an engineer at Chung-Hwa Telecommunication Company, Taiwan.

William S. Chao received his master degree (1985) in information science and Ph.D. degree (1988) in information science, both from the University of Alabama at Birmingham, USA. From 1988 till 1991, he worked as a computer scientist at GE Research and Development Center, Schenectady, New York, USA.

Dr. William S. Chao has been teaching at National Sun Yat-Sen University, Taiwan since 1992 and now serves as the president of Association of Enterprise Architects, Taiwan Chapter. His research covers: systems architecture, hardware architecture, software architecture, enterprise architecture, knowledge architecture and thinking architecture.

Dr. Shuh-Ping Sun was born in Taiwan. He received a Ph.D. degree in Mechanical Engineering from the Auburn University, AL, USA. He was an associate professor during 1995-2008, and has been a professor since Aug. 2008, in the Department of Biomedical Engineering at I-Shou University, Taiwan. He is a professor in the Department of Digital Media Design and the director of System Architecture research center at I-Shou University. He conducts research in Systems Architecture (SBC Architecture) and Creative Media Design.

PART I: BASIC CONCEPTS

Chapter 1: Why Does UML Need a Metamodel?

Unified Modeling Language (UML) 2.0 is a general modeling language for model-driven engineering (MDE) applications. The UML 2.0 specification defines a set of language concepts that is used to model the static structure and dynamic behavior of a system. The UML 2.0 concepts include (a) an abstract syntax that defines the language concepts and is described by a metamodel, and (b) a concrete syntax, or notation, that defines how the language concepts are represented and is described by a user model (or systems model).

Since UML is a multi-diagram approach, there are always some inconsistencies between different diagrams in the user model. To ensure and check the consistency, a metamodel that defines the abstract syntax of a modeling language needs to provide a unified semantic framework for defining consistency rules to impose constraints on the structure or behavior constructs. Hopefully through this unified semantic framework, each diagram in the user model can be projected as a view of the metamodel.

1-1 Unified Modeling Language

Unified modeling language (UML) [Blah04, Rumb91] is an object-oriented modeling language for model-driven (software) engineering (MDE) applications [Bram17, Weil09]. UML uses at least two types of diagrams, as shown in Figure 1-1, to represent different views of the system under consideration. (A) Structure type: emphasizes the static structure of the system using objects, attributes, operations and relationships. This structure type includes object diagram, class diagram, deployment diagram, package diagram, composite structure diagram, component diagram, etc. (B) Behavior type: emphasizes the dynamic behavior of the system by showing collaborations among objects and changes to the internal states of objects. This behavior type includes use case diagram, activity diagram, state diagram, sequence diagrams, communication diagram, interaction overview diagram, timing diagram, etc.

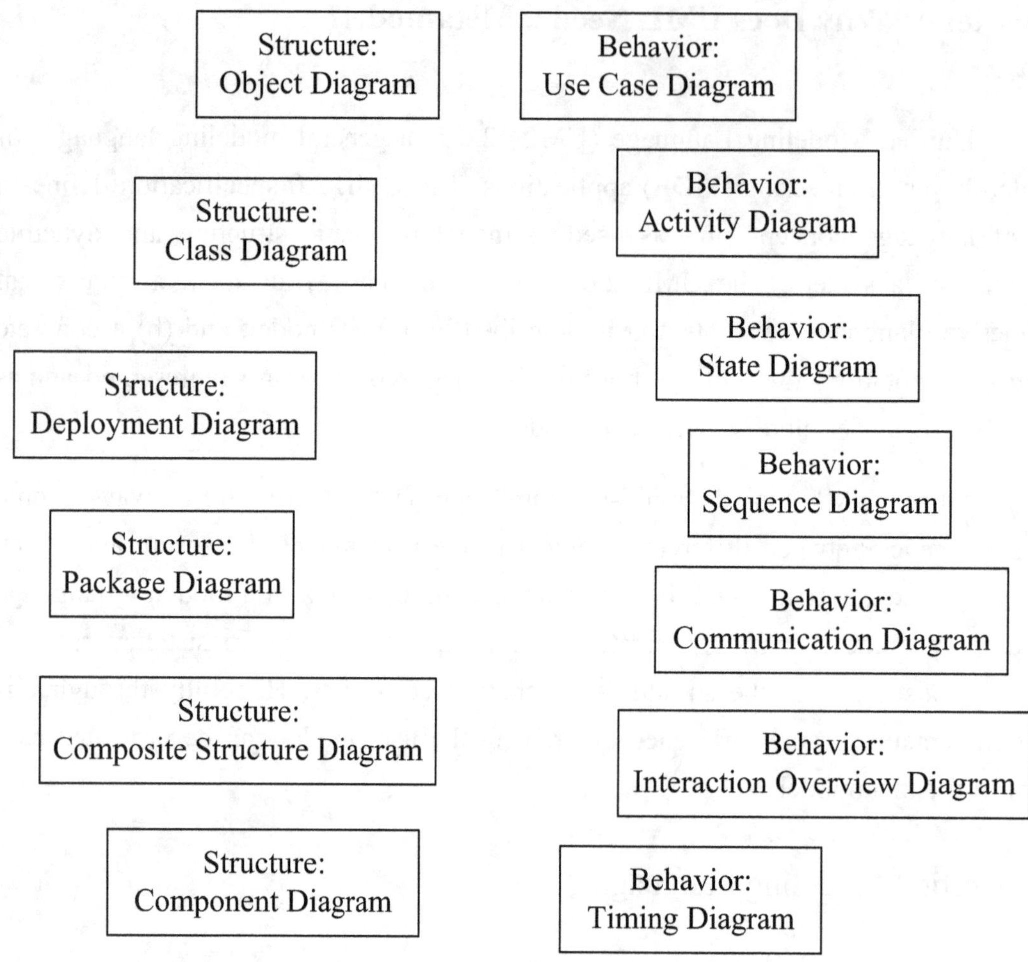

Figure 1-1. Multiple Diagrams of UML

1-2 Inconsistency Problems of UML

Since UML is a multi-diagram approach, there are some inconsistencies [Alla15, Dori95, Dori02, Dori16, Enge02, Malg06, Pele00, Przi16] between those different diagrams in the user model. In a multi-diagram environment, comprehending a system and the way it operates and changes over time requires concurrent reference to the various diagrams and the creation of abstract associations that link them. These multiple diagrams are separated. Rather than being built into the method, the mental burden of integrating the various diagrams is placed on the shoulder of the developers who need to deal with a system that is complex in itself, and they are mentally overloaded without any reason. Technical solutions that involve sophisticated tools can reduce manual consistency maintenance, but do not address the core issues of excessive mental burden.

In the UML multi-diagram environment, the straightforward intuition of thinking concurrently about the structure and behavior is seriously hindered by separating the structure and behavior diagrams. These multiple diagrams are dissociated and always contradictory with each other, which become the major cause for the UML inconsistency problems [Alla15, Bash16, Dori16].

1-3 Metamodel of UML

To ensure and check the consistency, we always need to create a kernel model for UML. This kernel model is the metamodel of UML. All UML structure views such as object diagram, class diagram and UML behavior views such as use case diagram, activity diagram, state diagram, sequence diagram, can be projected from this UML metamodel, as shown in Figure 1-2.

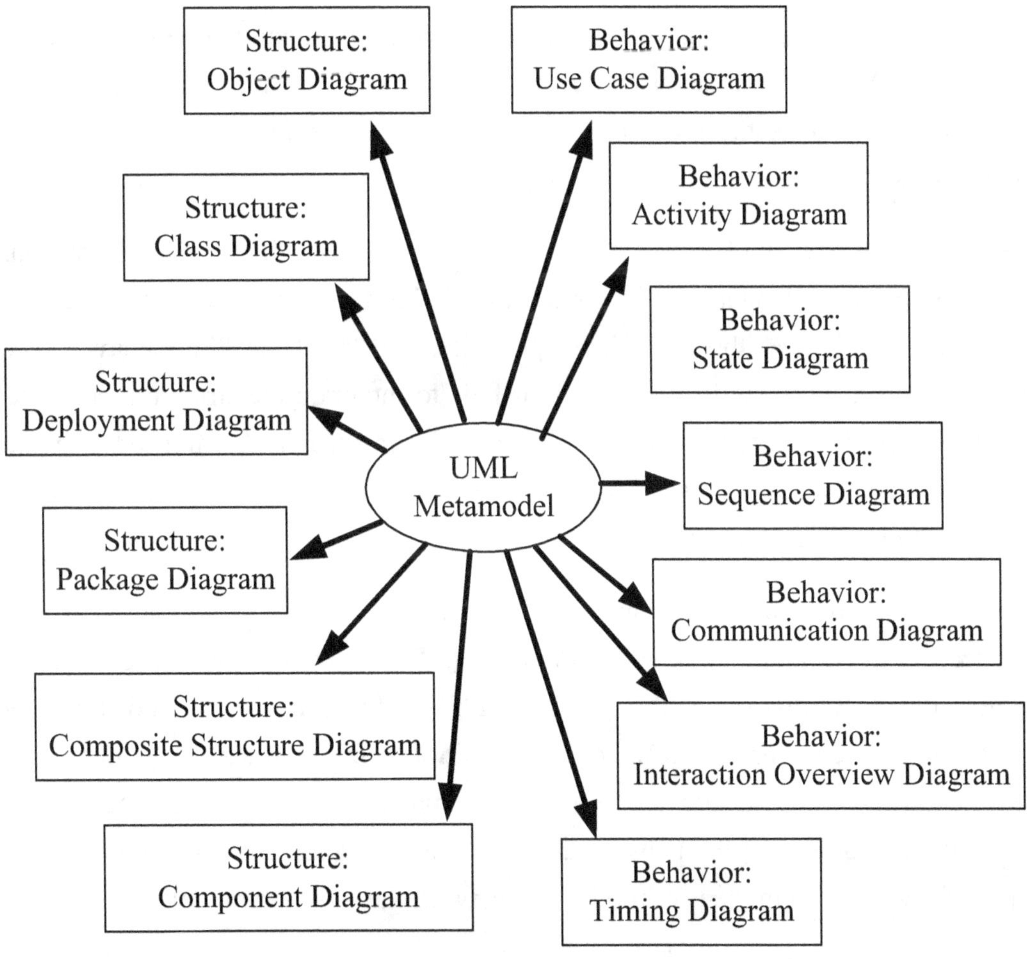

Figure 1-2. Metamodel of UML

1-4 Deficiency of Current UML Metamodels

In UML, a metamodel [Lale08] is used to describe the concepts in the language, their characteristics, and interrelationships. This is sometimes called the abstract syntax of the language, and is distinct from the concrete syntax that specifies the user model for the language. A significant usage of the metamodel is to ensure model consistency between different diagrams in the user model.

The Object Management Group (OMG) defines a language for representing metamodels, called Meta Object Facility (MOF) that is used to define UML, SysML and other metamodels. Several mechanisms are used in MOF, such as Object Constraint Language (OCL), Foundational UML (fUML), The Action Language for Foundational UML (Alf), Process Specification Language (PSL), to name a few.

The Object Constraint Language (OCL) is a precise text language that provides constraint on the structure (i.e., objects) to ensure consistency of the user model [Przi16]. However, not every diagram in the user model can be projected as a view of the OCL metamodel because the OCL fails to provide a unified semantic framework. Therefore, the OCL metamodel can only ensure part of (not all) user model consistency.

The Foundational UML is a subset of the standard UML for which a standard execution constraint language, PSL, is used to define the semantics of the execution model [OMG 13a]. Although fUML provides constraint on the behavior (i.e., activities) to make the model executable, it fails to integrate the structural constructs with the behavioral constructs. Not being able to provide a unified semantic framework, the Foundational UML can not project every diagram in the user model as a view of the fUML metamodel.

The Action Language for Foundational UML (Alf) is a complementary specification to Foundational UML [OMG 13b]. The key use of Alf is to act as the notation for specifying executable behaviors in UML, for example, methods for object operations, the behavior of an object, or transition effects on state machines. Like fUML, Alf also fails to provide a unified semantic framework to integrate the structural constructs with the behavioral constructs. Therefore, the Alf is not able to project every diagram in the user model as a view of the Alf metamodel.

1-5 SBC Approach for UML Metamodel

In order to overcome the shortcomings of the current UML metamodel approaches, we found that we need to develop a unified semantic framework that is able to integrate the structural constructs with the behavioral constructs. Operation-Based Multi-Queue Structure-Behavior Coalescence process algebra (O-M-SBC-PA) is such a candidate. In O-M-SBC-PA, the structural and behavioral constructs are integrated. Using O-M-SBC-PA as a metamodel for UML, each diagram in the user model can be projected as a view of the O-M-SBC-PA metamodel, as shown in Figure 1-3.

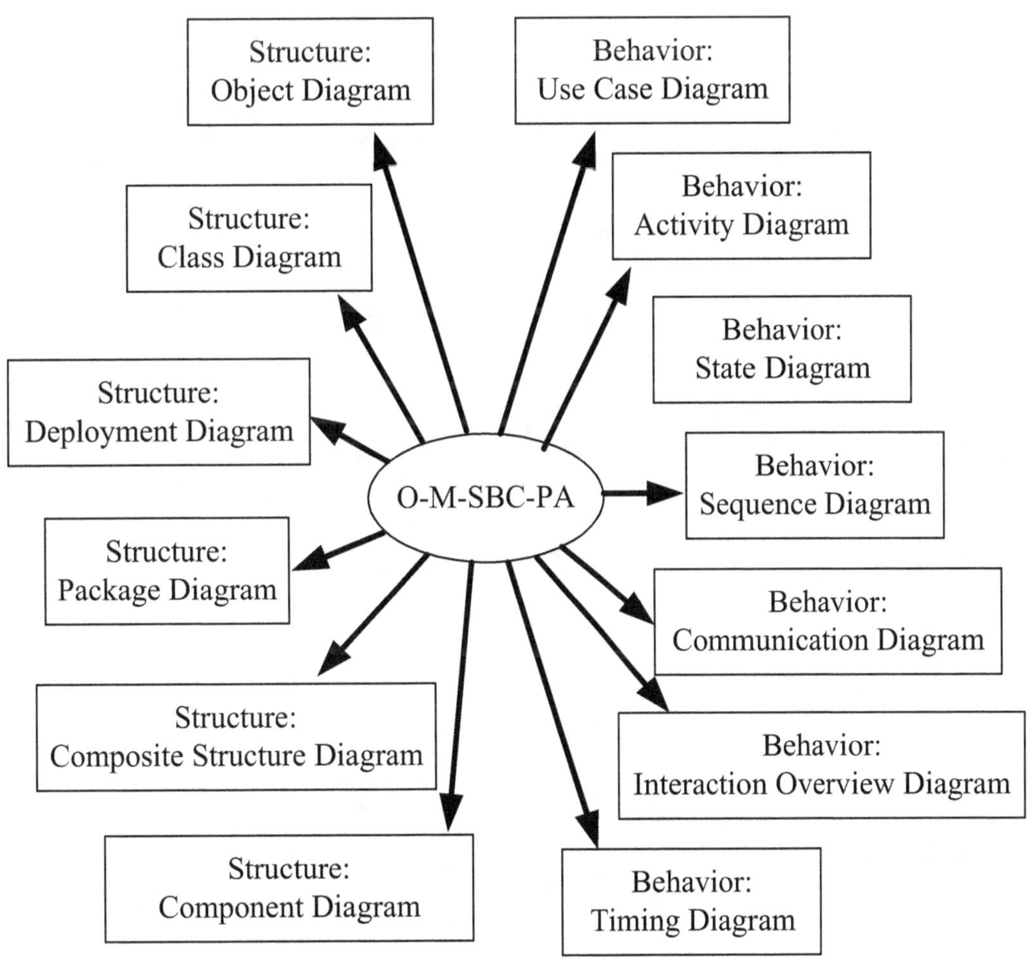

Figure 1-3. O-M-SBC-PA as the Metamodel of UML

Chapter 2: Introduction to Process Algebra

In this chapter, we first discuss what process algebra is. Then we show several examples of process algebras. After that we elaborate on the generalized SBC process algebra and six specialized SBC process algebras.

2-1 What is Process Algebra?

Process algebras are a diverse family of related approaches to the study of concurrent systems [Berg87, Chao15d, Hoar85, Miln89, Miln99]. Their tools are algebraic languages for the high-level description of interactions, communications and synchronizations between a collection of independent agents or processes.

Process algebras also provide algebraic laws that allow process descriptions to be manipulated and analyzed, and permit formal reasoning about equivalences and observation congruence among processes.

2-2 Well-known Examples of Process Algebra

There are several leading algebraic approaches to modeling concurrent systems.

Communicating Sequential Processes (CSP) [Hoar85] was first described in a 1978 paper by C. A. R. Hoare.

Arthur John Robin Gorell Milner introduced the Calculus of Communicating Systems (CCS) [Miln89, Miln99, Sang03] around 1980.

Algebra of Communicating Processes (ACP) [Berg87] was initially developed by Jan Bergstra and Jan Willem Klop in 1982.

2-3 Generalized SBC Process Algebra

Generalized SBC process algebra (G-SBC-PA) evolved from CCS (Calculus of Communicating Systems) [Miln89, Miln99].

CCS is a general process algebra language for the study of communication and concurrency. Like CCS, generalized SBC process algebra is also a general process algebra language for the study of communication and concurrency..

2-4 Specialized SBC Process Algebras

Channel-based single-queue SBC process algebra (C-S-SBC-PA) [Chao15d, Chao15e, Chao15g], channel-based multi-queue SBC process algebra (C-M-SBC-PA) [Chao15d, Chao15f, Chao15h], channel-based infinite-queue SBC process algebra (C-I-SBC-PA) [Chao15b, Chao15c, Chao15d], operation-based single-queue SBC process algebra (O-S-SBC-PA), operation-based multi-queue SBC process algebra (O-M-SBC-PA) [Chao15d, Chao15f, Chao15h] and operation-based infinite-queue SBC process algebra (O-I-SBC-PA) [Chao15b, Chao15c, Chao15d] are the six specialized SBC process algebras.

Channel-based single-queue SBC process algebra, channel-based multi-queue SBC process algebra, channel-based infinite-queue SBC process algebra, operation-based single-queue SBC process algebra, operation-based multi-queue SBC process algebra and operation-based infinite-queue SBC process algebra all evolved from CCS (Calculus of Communicating Systems) [Miln89, Miln99].

CCS is a general process algebra language for the study of concurrent systems. Unlike CCS, six specialized SBC process algebras are only applicable to systems model [Burd10, Maie09, Chao16b, Chao16c, Chao16d, Chao16e, Chao16f, Chao16g, Craw15, Chec99, Dam06, O'Rou03, Putm00, Rayn09, Roza11, Toga08].

Chapter 3: Mathematics of Process Algebra

To give the process algebra a mathematical definition, we need a means to form new processes from old ones. The basic operators, always present in some form or other, allow sequentialization of interactions or summation of processes or parallel composition of processes or recursive definition of a process or replication of processes or conditional definition of a process or null process.

3-1 Sequentialization of Interactions

Sometimes interactions must be temporally ordered. For example, it might be desirable to specify algorithms such as: execute the "a" interaction first and then execute the "*P*" process later. Sequentialization of interactions can be used for such purposes.

Sequentialization of interactions, usually written as the $a \bullet P$ process, indicates that it will perform the "a" interaction first and continue as the P process.

3-2 Summation of Processes

The binary operator "+", summation, combines two process expressions as alternatives.

For example, the P_1+P_2 process can advance non-deterministically either as the P_1 process or the P_2 process; as soon as one performs its first interaction the other is discarded.

3-3 Parallel Composition of Processes

Parallel composition of two processes P_1 and P_2, usually written $P_1 \square P_2$, is the key primitive distinguishing the process algebras from sequential models of process executions.

Parallel composition allows the executions in P_1 and P_2 to proceed simultaneously and independently.

3-4 Recursive Definition of a Process

The operators presented so far describe only finite interaction and are consequently insufficient for full computability, which includes non-terminating behavior. Recursion is the operator that allows finite descriptions of infinite behavior.

For example, **fix**(X=E) can be understood as abbreviating the recursive definition of an infinite behavior denoted by the "X" process variable.

3-5 Replication of a Process

Replication is the other operator that allows finite descriptions of infinite behavior of a process.

For example, replication **!**P can be understood as abbreviating the parallel composition of a countably infinite number of P processes.

3-6 Conditional Definition of a Process

A process can be defined by a one-or-more-armed conditional expression. For example, the process (**if** $cond_1$ **then** P_1)+(**if** $cond_2$ **then** P_2)…+(**if** $cond_j$ **then** P_j) will proceed as the process P_1 if the "$cond_1$" value is true, or proceed as the process P_2 if the "$cond_2$" value is true,…, or proceed as the process P_j if the "$cond_j$" value is true.

3-7 Null Process

Process algebras generally also include a null process, denoted as *STOP*, which has no interaction points. It is utterly inactive and its sole purpose is to act as the inductive anchor on top of which more interesting processes can be generated.

The process "$STOP \bullet P_1$" (i.e. sequential composition of processes *STOP* and P_1) equals to the process "*STOP*", as shown in Figure 3-1.

$$STOP \bullet P_1 \quad = \quad STOP$$

Figure 3-1. Characteristics of the Null Process (I)

The process "P_2+*STOP*" (i.e. summation of processes P_2 and *STOP*) equals to the process "*STOP*+P_2" (i.e. summation of processes *STOP* and P_2) which equals to the process "P_2", as shown in Figure 3-2.

$$P_2 + STOP \quad = \quad STOP + P_2 \; = \; P_2$$

Figure 3-2. Characteristics of the Null Process (II)

The process "$P_3 \square STOP$" (i.e. parallel composition of processes P_3 and *STOP*) equals to the process "$STOP \square P_3$" (i.e. parallel composition of processes *STOP* and P_3) which equals to the process "P_3", as shown in Figure 3-3.

$$P_3 \square STOP \quad = \quad STOP \square P_3 \quad = \quad P_3$$

Figure 3-3. Characteristics of Null Process (III)

PART II: OPERATION-BASED MULTI-QUEUE SBC PROCESS ALGEBRA

Chapter 4: Operation-Based Value-Passing Interactions

In this chapter, we first introduce operations and operation-based interactions. We then introduce the formal description of an operation-based communication port, an operation-based action and an operation-based interaction.

4-1 Operations

An operation provided by each object represents a procedure, or method, or function of the object as shown in Figure 4-1.

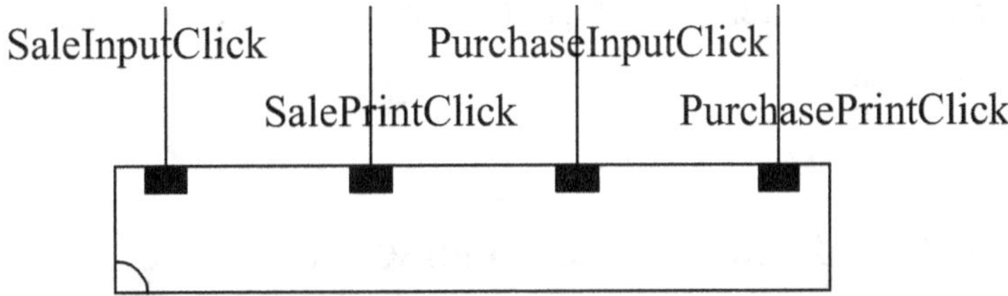

Figure 4-1. An Operation Represents a Procedure, or Method, or Function of an Object

An operation may contain several input parameters (e.g. i_1, i_2) and output parameters (e.g. o_1, o_2), as shown in Figure 4-2.

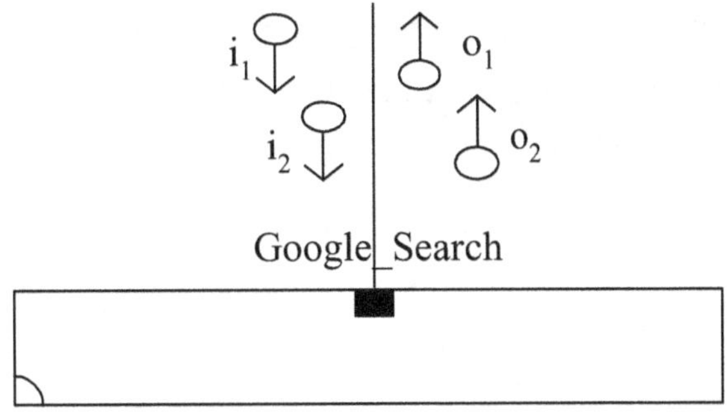

Figure 4-2. An Operation Contains Several Input/Output Parameters

An operation (which can be extended to operation call or operation return) signature is used to completely describe an operation. The signature for an operation is a combination of its name along with parameters as follows:

<operation name> (<parameter list>)

The parameters in the parameter list represent the inputs or outputs of the operation. Each parameter in the list is displayed with the following format:

<direction> <parameter name> : <parameter type>

Parameter direction may be in, out, or inout. We formally describe the "operation signature" as a relation $L \subseteq \Lambda X \Theta$ where Λ is a set of "operation names" and Θ is a set of "parameter lists".

4-2 Operation-Based Value-Passed Interactions

An interaction represents an indivisible and instantaneous handshake or communication between two agents. In the operation-based value-passing approach as shown in Figure 4-3, the caller agent (either external environment's actor or object) communicates with the callee agent (object) through the operation call (solid line) or operation return (dashed line) interaction. In the figure, "Google_Search (in a; in b)" is an operation call signature and "Google_Search (out c; out d)" is an operation return signature. The operation call signature and its corresponding operation return signature can be merged into an operation signature (i.e., Google_Search (in a; in b; out c; out d). The figure also depicts that the "Google_Search" operation is **required** by the caller agent and is **provided** by the callee agent.

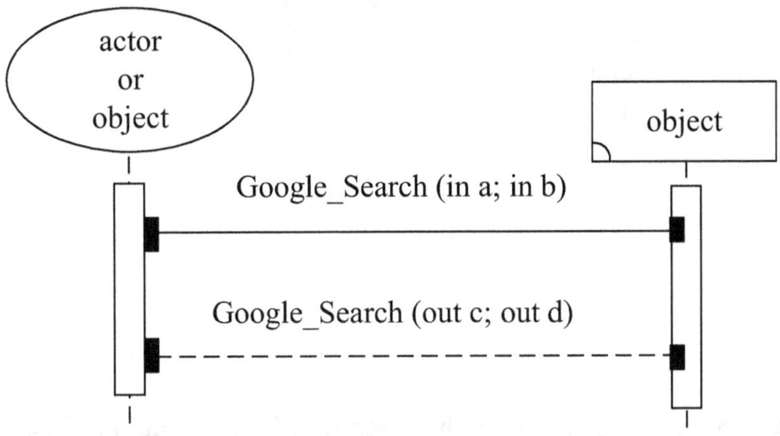

Figure 4-3. Operation-Based Value-Passing Interactions

The caller agent owns the "calling port" of the interaction. In the "operation call interaction" case, the calling port is "Google_Search (in a; in b)" whose conduct is to output a value to each one of the "a" and "b" variables (of the "Google_Search" operation) as shown in Figure 4-4.

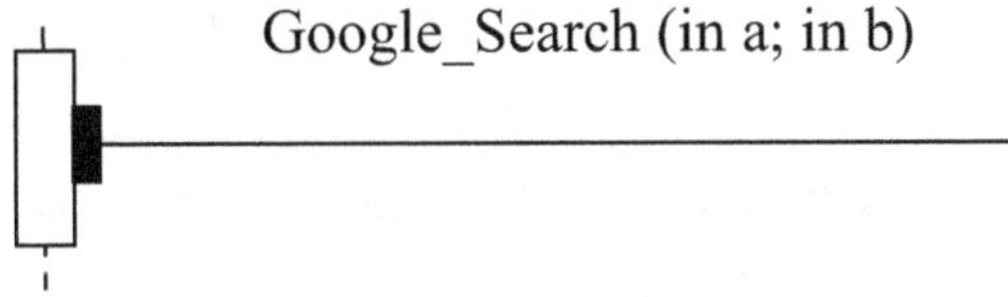

Figure 4-4. Calling Port in the Operation Call Interaction Case

The caller agent together with the "calling port" is named the "calling action" as shown in Figure 4-5.

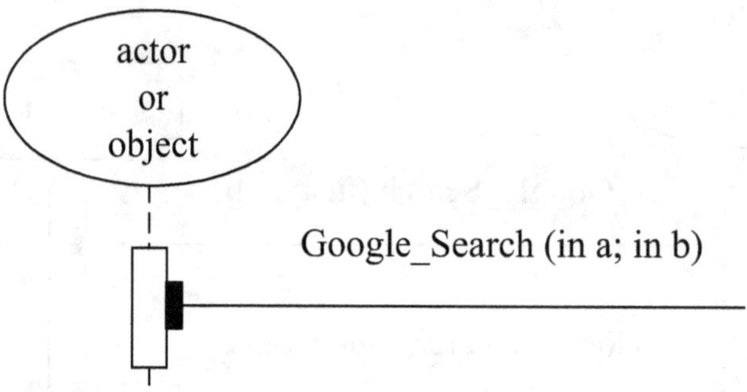

Figure 4-5. Calling Action in the Operation Call Interaction Case

In the "operation return interaction" case, the calling port is "Google_Search (out c; out d)" whose conduct is to receive a value from each one of the "c" and "d" variables (of the "Google_Search" operation) as shown in Figure 4-6.

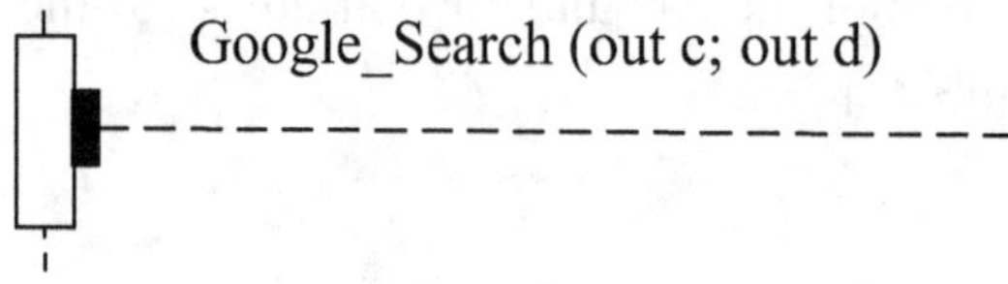

Figure 4-6. Calling Port in the Operation Return Interaction Case

The caller agent together with the "calling port" is named the "calling action" as shown in Figure 4-7.

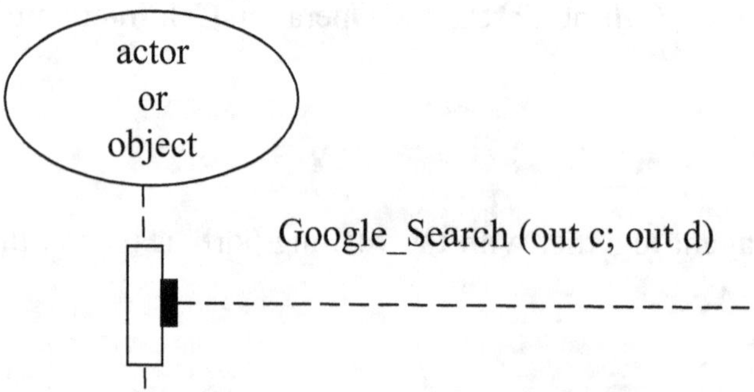

Figure 4-7. Calling Action in the Operation Return Interaction Case

The callee agent owns the "called port" of the interaction. In the "operation call interaction" case, the called port is "Google_Search (in a; in b)" whose conduct is (using the "Google_Search" operation) to receive a value for each one of the "a" and "b" variables as shown in Figure 4-8.

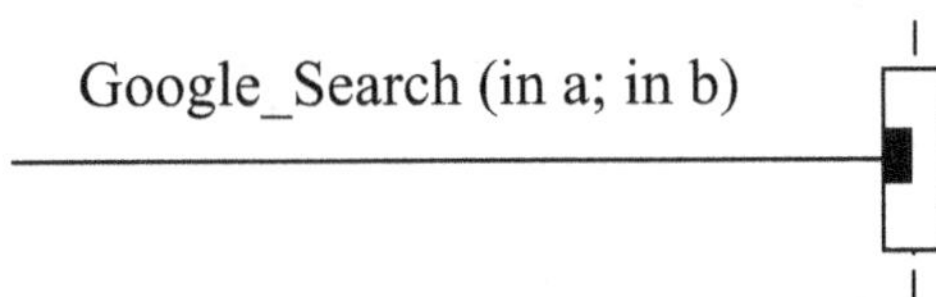

Figure 4-8. Called Port in the Operation Call Interaction Case

The callee agent together with the "called port" is named the "called action" as shown in Figure 4-9.

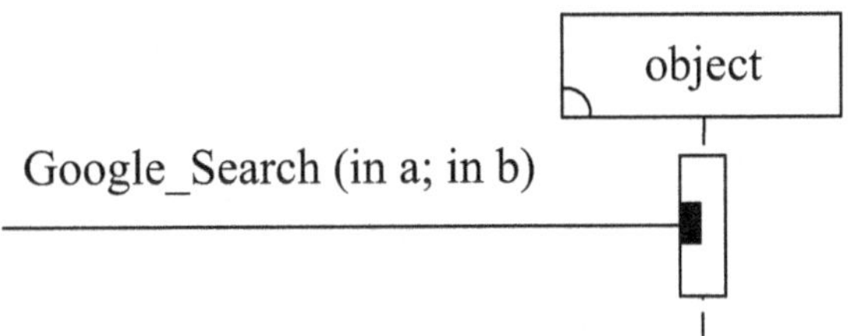

Figure 4-9. Called Action in the Operation Call Interaction Case

In the "operation return interaction" case, the called port is "Google_Search (out c; out d)" whose conduct is (using the "Google_Search" operation) to output the values of the "c" and "d" variables as shown in Figure 4-10.

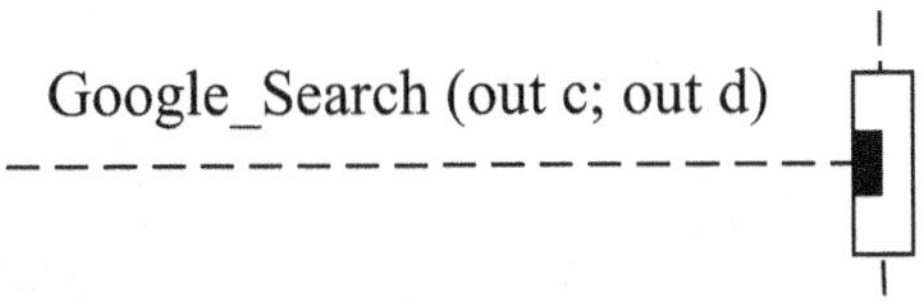

Google_Search (out c; out d)

Figure 4-10. Called Port in the Operation Return Interaction Case

The callee agent together with the “called port” is named the “called action” as shown in Figure 4-11.

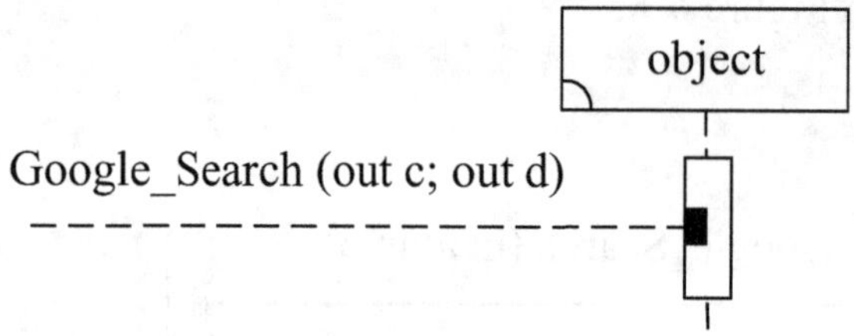

Figure 4-11. Called Action in the Operation Return Interaction Case

In order to simplify the “operation-based value-passed interaction” diagram, we will redraw it as shown in Figure 4-12.

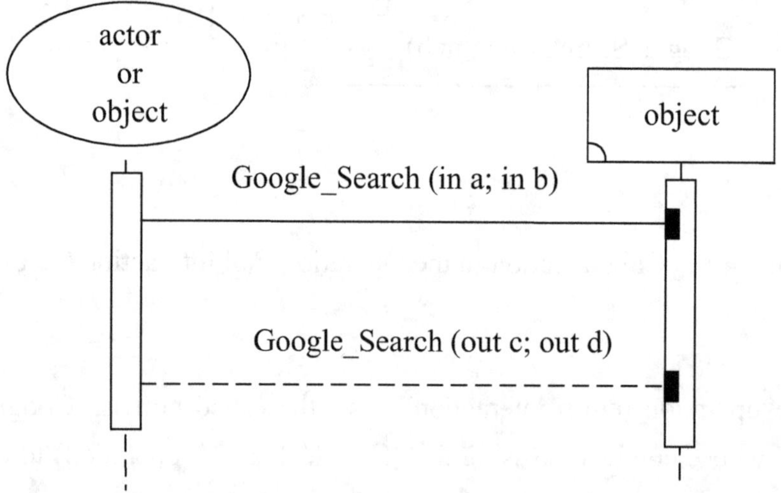

Figure 4-12. Operation-Based Interaction Value-Passing Diagram (I)

Or we can draw the operation-based value-passing interaction diagram as shown in Figure 4-13.

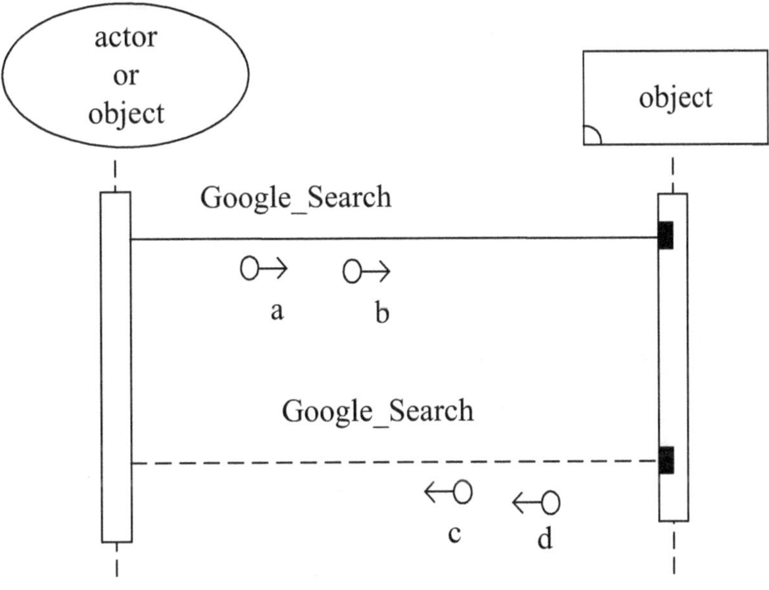

Figure 4-13. Operation-Based Value-Passing Interaction Diagram (II)

4-3 Formal Description of an Operation-Based Interaction

The external environment uses a "type 1 interaction" to interact with an object. We formally describe the operation-based value-passing "type 1 interaction" as a relation $G \subseteq NXAXLX\Gamma$ where N is a set of "operation call or operation return tags" and A is a set of "external environment's actors" and L is a relation of "operation call or operation return signatures" and Γ is a set of "objects".

Two objects use a "type 2 interaction" to interact with each other. We formally describe the operation-based value-passing "type 2 interaction" as a relation $V \subseteq NX\Gamma XLX\Gamma$ where N is a set of "operation call or operation return tags" and Γ is a set of "objects" and L is a relation of "operation call or operation return signatures".

We can also formally describe the operation-based value-passing "type 1 or 2 interaction" as a relation $\Delta \subseteq NX\Xi XLX\Gamma$ where N is a set of "operation call or operation return tags" and Ξ is a set of "external environment's actors or objects" and L is a relation of "operation call or operation return signatures" and Γ is a set of "objects".

Chapter 5: The Structure-Behavior Coalescence Approach

Software structure and software behavior are the two most prominent views of a software system, integrating the software structure and software behavior is apparently the best way to achieve an integrated whole of a software system. If we are not able to integrate the software structure and software behavior, then there is no way that we are able to integrate the whole software system. Structure-behavior coalescence (SBC) provides an elegant way to integrate the software structure and software behavior of a system. In other words, SBC facilitates an integrated whole of a software system.

5-1 Structure-Behavior Coalescence Means to Integrate the Software Structure and Software Behavior

Software structure, specified by objects, their operations and their composition, refers to the type of connection between the objects of a software system as shown in Figure 5-1.

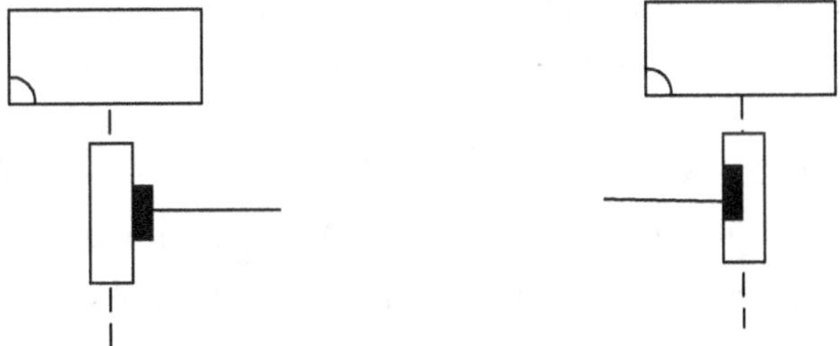

Figure 5-1. Software Structure

Software behavior, specified by the interactions between and among the objects and environment, refers to the interconnectivities a software system in conjunction with its environment as shown in Figure 5-2.

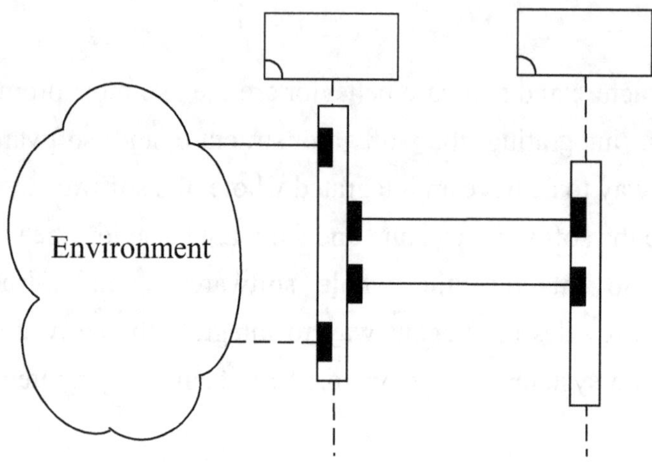

Figure 5-2. Software Behavior

Software structure and software behavior are the two most prominent views of a software system, integrating the software structure and software behavior is apparently the best way to achieve an integrated whole of a software system.

If we are not able to integrate the software structure and software behavior, then there is no way that we are able to integrate the whole software system.

Structure-behavior coalescence (SBC) [Chao15a, Lin19] provides an elegant way to integrate the software structure and software behavior of a software system. In other words, SBC facilitates an integrated whole of a software system.

5-2 Interactions among Objects and the External Environment to Draw Forth the Software Behavior

All things that strike us as something independent are essentially parts of a software system. We usually call the parts of a software system its objects. Objects are sometimes labeled as parts, entities, components, building blocks and non-aggregated systems [Chao14a, Chao14b, Chao14c, Chao16a].

In a software system, if the objects, and among them and the external environment to interact (or handshake), such interaction will draw forth the software behavior.

The external environment uses a “type-1 interaction” to interact with an object as shown in Figure 5-3.

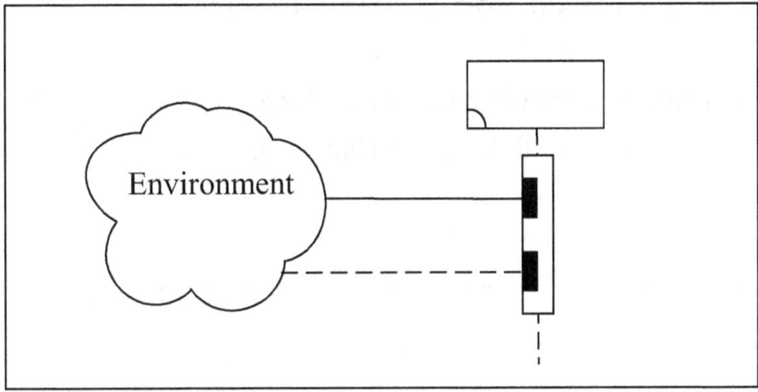

Figure 5-3. The External Environment Uses a "Type-1 Interction" to Interact with an Object

We formally describe the operation-based "type 1 interaction" as a relation $G \subseteq N X B X L X \Gamma$ where N is a set of "operation call or operation return tags" and B is a set of "external environment's actors" and L is a relation of "operation call or operation return signatures" and Γ is a set of "objects".

Two objects use a "type-2 interaction" to interact with each other as shown in Figure 5-4.

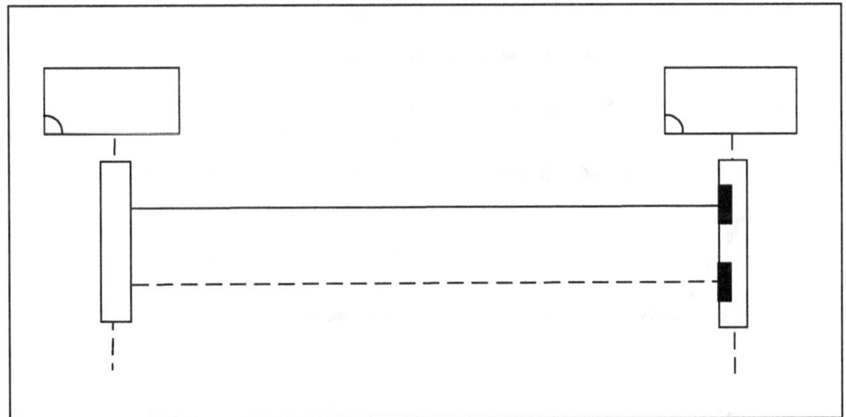

Figure 5-4. Two Objects Use a "Type-2 Interaction" to Interact with Each Other

We formally describe the operation-based "type 2 interaction" as a relation $V \subseteq N X \Gamma X L X \Gamma$ where N is a set of "operation call or operation return tags" and Γ is a set of "objects" and L is a set of "operation call or operation return signatures".

5-3 Core Theme of Structure-Behavior Coalescence

The core theme of structure-behavior coalescence is: "Software Architecture = Software Structure + Software Behavior." That is, the software structure will lead to the software behavior as shown in Figure 5-5.

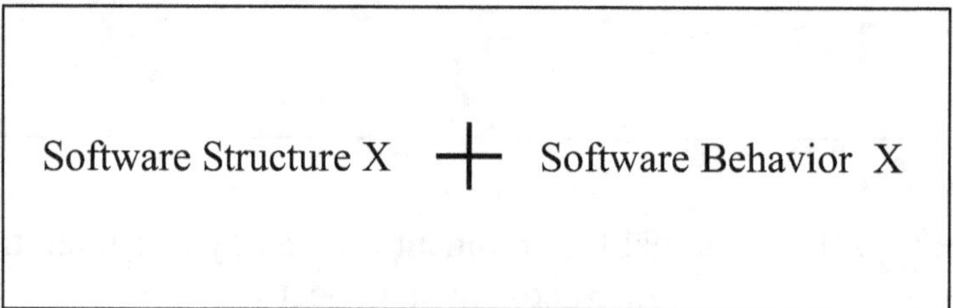

Figure 5-5. Core Theme of Structure-Behavior Coalescence

One software structure will draw forth one software behavior. That is, the software behavior is attached to or built on the software structure in the SBC approach.

In other words, the software behavior can not exist alone; it must be loaded on the software structure just like a cargo is loaded on a ship as shown in Figure 5-6.

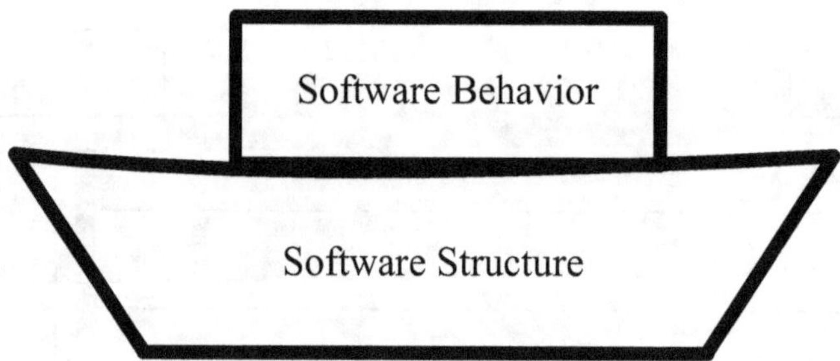

Figure 5-6. Software Behavior Must be Loaded on the Software Structure

Chapter 6: Language Constructs of Operation-Based Multi-Queue SBC Process Algebra

In the chapter, we illustrate in detail those operation-based multi-queue SBC process algebra language constructs which make up the metamodel of UML.

6-1 Entity Set and Entity Name

As shown in Figure 6-1, we assume a relation L of operation call or operation return signatures, and use l_1, l_2... to range over L. Further, we let Λ be the set of operation names, and use op_1, op_2... to range over Λ. We let Θ be the set of parameter lists, and use p_1, p_2... to range over Θ. We let G be the relation of type 1 interactions, and use g_1, g_2... to range over G. We let V be the relation of type 2 interactions, and use v_1, v_2... to range over V. We let Δ be the relation of type 1 or 2 interactions, and use a_1, a_2... to range over Δ. We let Π be the set of processes, and use P_1, P_2... to range over Π. We let Ψ be the set of process expressions, and use E_1, E_2... to range over Ψ. Further, we let X be the set of process variables, and use X_1, X_2...to range over X. We let Φ be the set of process Constants, and use A_1, A_2... to range over Φ. We let B be the set of actors, and use β_1, β_2... to range over B. We let Γ be the set of objects, and use O_1, O_2... to range over Γ. We let Ξ be the set of actors or objects, and use ρ_1, ρ_2... to range over Ξ. Finally, we let N be the set of operation call or operation return tags, and use n_1, n_2... to range over N.

Entity set or relation	Entity name	Type of entity
L	$l_1, l_2...$	operation call or operation return signatures
Λ	op_1, op_2	operation names
Θ	$p_1, p_2...$	parameter lists
G	$g_1, g_2...$	type 1 interactions
V	$v_1, v_2...$	type 2 interactions
Δ	$a_1, a_2...$	type 1 or 2 interactions
	I, J,...	indexing sets
Π	$P_1, P_2...$	processes
Ψ	$E_1, E_2...$	process expressions
X	$X_1, X_2...$	process variables
Φ	$A_1, A_2...$	process Constants
B	$\beta_1, \beta_2...$	actors
Γ	$o_1, o_2...$	objects
Ξ	$\rho_1, \rho_2...$	actors or objects
N	$n_1, n_2...$	operation call or operation return tag

Figure 6-1. Entities of Operation-Based Multi-Queue SBC Process Algebra

6-2 Backus-Naur Form of Operation-Based Multi-Queue SBC Processes

The set of operation-based multi-queue SBC processes is defined by the following BNF grammar, as shown in Figure 6-2.

(1) <System> ::= <FixIFD> {" ▯ " <FixIFD>}

(2) <FixIFD> ::= "**fix(**" <Process_Variable>"="<IFD>
"●" <Process_Variable> ")"

(3) <IFD> ::= <Type_1_Interaction> {"● " Type_1_Or_2_Interaction>}

(4) <Type_1_Or_2_Interaction> ::= <Type_1_Interaction>

| <Type_2_Interaction>

Figure 6-2. Backus-Naur Form of
Operation-Based Multi-Queue SBC Processes

6-2-1 Parallcl Composition of One or More Recursive Interaction Flow Diagrams Defines the Operation-Based Multi-Queue SBC Process of a System

Rule 1 describes that parallel composition of one or more recursive interaction flow diagrams (i.e. FixIFD) defines the operation-based multi-queue SBC process of a system, as shown in Figure 6-3.

Rule 1
<System> ::= <FixIFD> {" ▯ " <FixIFD>}

Figure 6-3. Rule 1

6-2-2 A Recursive Interaction Flow Diagram is the Recursion of an Interaction Flow Diagram

Rule 2 describes that a recursive interaction flow diagram (i.e. FixIFD) is defined by the recursion of an interaction flow diagram (i.e. IFD), as shown in Figure 6-4.

Rule 2
<FixIFD> ::= "**fix(**" <Process_Variable> "=" <IFD> " • " <Process_Variable> ")"

Figure 6-4. Rule 2

6-2-3 An Interaction Flow Diagram is a Type_1 Interaction Followed by Zero or More Type_1_Or_Type_2 Interactions

Rule 3 describes that an interaction flow diagram (i.e. IFD) is defined by a type_1 interaction (i.e. Type_1_Interaction) followed by zero or more type_1_or_type_2 interactions (i.e. Type_1_Or_2_Interaction), as shown in Figure 6-5.

Rule 3
<IFD> ::= <Type_1_Interaction> {"• " <Type_1_Or_2_Interaction>}

Figure 6-5. Rule 3

6-2-4 Type_1_Or_2 Interaction is either Type_1 or Type_2

Rule 4 describes that the type_1_or_2 interaction (i.e. Type_1_Or_2_Interaction) is either a type_1 interaction (i.e. Type_1_Interaction) or a type_2 interaction (i.e. Type_2_Interaction), as shown in Figure 6-6.

Rule 4		
<Type_1_Or_2_Interaction>	::=	<Type_1_Interaction>
	|	<Type_2_Interaction>

Figure 6-6. Rule 4

Chapter 7: Transitional Semantics of Operation-Based Multi-Queue SBC Process Algebra

In the chapter, we illustrate in detail those operation-based multi-queue SBC process algebra transitional semantics which describes the metamodel of UML.

7-1 Transitional Semantics

In giving meaning to the operation-based multi-queue SBC process algebra, we shall use the following labelled transition system (LTS) [Miln89, Miln99].

$$(\Psi, \Delta, \rightarrow)$$

which consists of a set Ψ of process expressions (process expressions are identified with states), a set Δ of "type 1 or 2 interactions", and a transition relation $\rightarrow \subseteq \Psi_1 X \Delta X \Psi_2$ where $(E_j, \mathrm{a}, E_k) \in \rightarrow$ is denoted by $E_j \xrightarrow{\mathrm{a}} E_k$.

The semantics for Ψ consists in the transition rules of each transition relation $\rightarrow$ over $\Psi_1 X \Delta X \Psi_2$. These transition rules will follow the construct of process expressions.

As shown in Figure 7-1, we give the complete set of transition rules; the names Prefix, Sum, Recursion, Parallel, and Constant indicate that the rules are associated respectively with Prefix, Summation, Recursion, and Parallel Composition and with Constants.

Prefix	$\dfrac{}{a \bullet E \xrightarrow{a} E}$
Sum_j	$\dfrac{E_j \xrightarrow{a} E_j'}{\sum_{i \in I} E_i \xrightarrow{a} E_j'} \; (j \in I)$
Recursion	$\dfrac{E\{\mathbf{fix}(X=E)/X\} \xrightarrow{a} E'}{\mathbf{fix}(X=E) \xrightarrow{a} E'}$
Parallel_1	$\dfrac{E \xrightarrow{a} E'}{E \,\|\, F \xrightarrow{a} E' \,\|\, F}$
Parallel_2	$\dfrac{F \xrightarrow{a} F'}{E \,\|\, F \xrightarrow{a} E \,\|\, F'}$
Constant	$\dfrac{P \xrightarrow{a} P'}{A \xrightarrow{a} P'} \; (A \overset{\text{def}}{=} P)$

Figure 7-1. Transition Rules for the
Operation-Based Multi-Queue SBC Process Algebra

7-2 Rule of Prefix

The rule for Prefix, shown in Figure 7-2, can be read as follows: Under any circumstances, we always infer $a \bullet E \xrightarrow{a} E$. That is, an expression, with an interaction prefixed to it, will use this interaction to accomplish the transition.

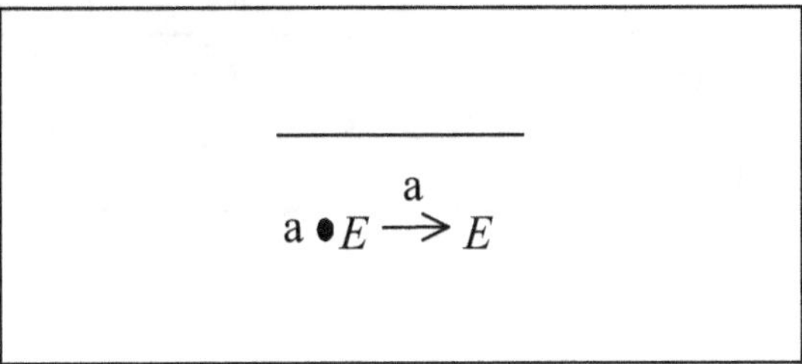

Figure 7-2. Rule of Prefix

7-3 Rule of Summation

The rule for Summation, shown in Figure 7-3, can be read as follows: If any one summand E_j of the sum $\sum_{i \in I} E_i$ has an interaction, then the whole sum also has that interaction.

$$\frac{E_j \xrightarrow{a} E_j'}{\sum_{i \in I} E_i \xrightarrow{a} E_j'} \quad (j \in I)$$

Figure 7-3. Rule of Summation

7-4 Rule of Recursion

The rule for Recursion, shown in Figure 7-4, can be read as follows: This says that any interaction which may be inferred for the **fix** expression 'unwound' once (by substituting itself for its bound variable) may be inferred for the **fix** expression itself.

$$\frac{E\{\mathbf{fix}(X{=}E)/X\} \xrightarrow{a} E'}{\mathbf{fix}(X{=}E) \xrightarrow{a} E'}$$

Figure 7-4. Rule of Recursion

7-5 Rule of Parallel Composition

There are two transition rules for parallel composition. Rule Parallel_1, as shown in Figure 7-5, indicates that from $E \xrightarrow{a} E'$ we shall infer $E \| F \xrightarrow{a} E' \| F$.

$$\frac{E \xrightarrow{a} E'}{E \| F \xrightarrow{a} E' \| F}$$

Figure 7-5. Rule Parallel_1

Rule Parallel_2, as shown in Figure 7-6, indicates that from $F \xrightarrow{a} F'$ we shall infer $E \| F \xrightarrow{a} E \| F'$.

$$\frac{F \xrightarrow{a} F'}{E \| F \xrightarrow{a} E \| F'}$$

Figure 7-6. Rule Parallel_2

7-6 Rule of Constants

The rule for Constants, shown in Figure 7-7, can be read as follows: the rule of Constants asserts that each Constant has the same transitions as its defining expression.

$$\frac{P \xrightarrow{a} P'}{A \xrightarrow{a} P'} \quad (A \stackrel{\text{def}}{=} P)$$

Figure 7-7. Rule of Constants

PART III: MAPPING FROM O-M-SBC-PA TO UML

Chapter 8: O-M-SBC-PA Transition Graph

In the chapter, we illustrate in detail those operation-based multi-queue SBC process algebra transition graphs which describe the semantics of the metamodel of UML.

8-1 Definition of a System in O-M-SBC-PA

In operation-based multi-queue SBC process algebra, the process expression of a system is defined as $\square_{i=1,m} \text{FixIFD}_i$ and the process expression of FixIFD_i is defined as $\mathbf{fix}(X_i = \bullet_{j=1,n} a_{ij} \bullet X_i)$, where $a_{i1}=g_{i1}$ for all $i \in I$. To combine them together, we summarize that in O-M-SBC-PA a system is then formally defined as "$\mathbf{fix}(X_1=g_{11}\bullet a_{12}\bullet a_{13}\bullet\ldots\bullet a_{1n}\bullet X_1)\square\mathbf{fix}(X_2=g_{21}\bullet a_{22}\bullet a_{23}\bullet\ldots\bullet a_{2n}\bullet X_2)\square \ldots\square\mathbf{fix}(X_m=g_{m1}\bullet a_{m2}\bullet a_{m3}\bullet\ldots\bullet a_{mn}\bullet X_m)$".

8-2 Transition Graphs in O-M-SBC-PA

Based on the O-M-SBC-PA transitional semantics, whenever $E_{01} \xrightarrow{a_{11}} \cdots \xrightarrow{a_{mn}} E_{01}'$, we call $(a_{11}\ldots a_{mn}, E_{01}')$ a derivative of E_{01}. It is convenient to collect the derivatives of a process expression E_{01} into the O-M-SBC-PA transition graph (TG) of E_{01}. We use the O-M-SBC-PA transition graph TG_{01} to define the execution of the process expression E_{01}, as shown in Figure 8-1.

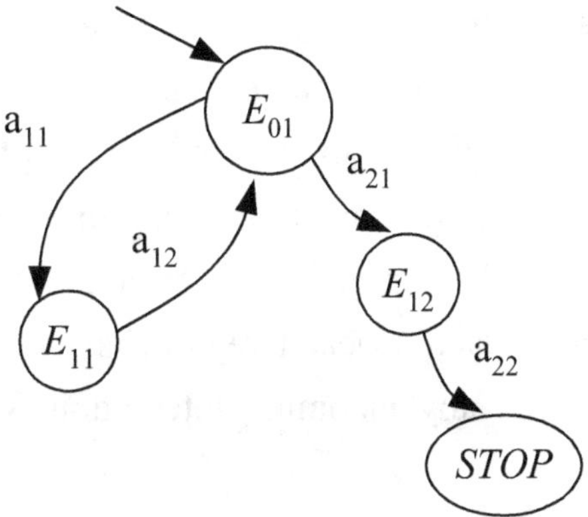

Figure 8-1. Transition Graph "TG_{01}" of the Process Expression "E_{01}"

In the O-M-SBC-PA transition graph of the process expression "E_{01}", the process expression (process expression is identified with state) is represented by a labeled circle; the edge is used to represent the "transition" between the two process expressions; the starting process expression is usually represented by an arrow with no origin pointing to the process expression.

8-3 TG Relations (TGR) in O-M-SBC-PA

We use a transition relation $TGR_{01} \subseteq \Psi_1 X \Delta X \Psi_2$, where Ψ is a set of "process expressions" and Δ is a relation of "type 1 or 2 interactions" and $(E_j, a, E_k) \in TGR_{01}$ is denoted by $E_j \xrightarrow{a} E_k$, to represent the O-M-SBC-PA transition graph TG_{01} of the process expression E_{01}, as shown in Figure 8-2.

Ψ_1	Δ	Ψ_2
E_{01}	a_{11}	E_{11}
E_{11}	a_{12}	E_{01}
E_{01}	a_{21}	E_{12}
E_{12}	a_{22}	*STOP*

Figure 8-2. Relation "TGR_{01}" of the Transition Graph "TG_{01}"

8-4 Orthogonal Composite State in the Transition Graph

When a process expression involves parallel composition rules, its corresponding transition graph may become complicated. In order to reduce the complexity of the transition graph, we shall introduce an orthogonal composite state. An orthogonal composite state in the transition graph may have many regions, which may each contain substates. These regions are orthogonal to each other. When an orthogonal composite state is active, each region has its own active state that is independent of the others and any incoming interaction is independently analyzed within each region.

An orthogonal composite state in the transition graph is represented as $\|_{i \in I} TG_i$ if each process expression "E_i" has a corresponding transition graph

"TG_i". Let us use an example to illustrate how the orthogonal composite state reduces the complexity of the transition graph. We define the process expressions "E_3".as "$a_3 \bullet a_4 \bullet STOP$" and "$E_5$".as "$a_5 \bullet a_6 \bullet STOP$". We use the O-M-SBC-PA transition graphs "TG_3" and "TG_5" to respectively define the execution of the process expressions "E_3" and "E_5", as shown in Figure 8-3.

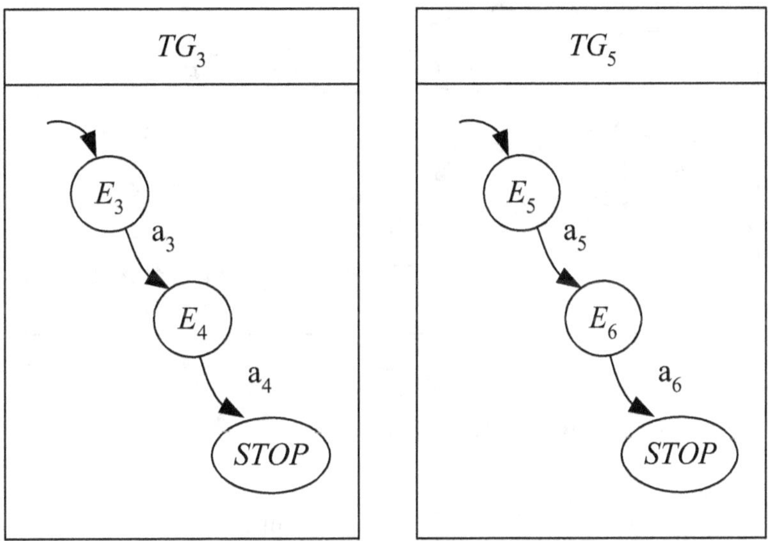

Figure 8-3. Transition Graphs TG_3 and TG_5

The process expression "E_7".is defined as "$a_3 \bullet a_4 \bullet STOP \square a_5 \bullet a_6 \bullet STOP$". There are two ways to define the O-M-SBC-PA transition graph for "E_7". The first one does not use the orthogonal composite state and results in the transition graph "TG_{71}", as shown in Figure 8-4.

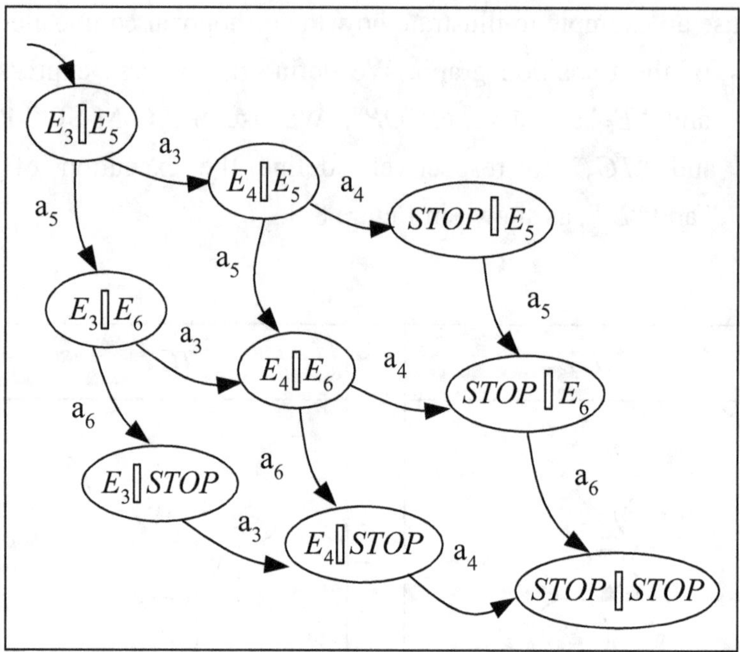

Figure 8-4. Transition Graph TG_{71}

The second one uses the orthogonal composite state and results in the transition graph "TG_{72}", as shown in Figure 8-5.

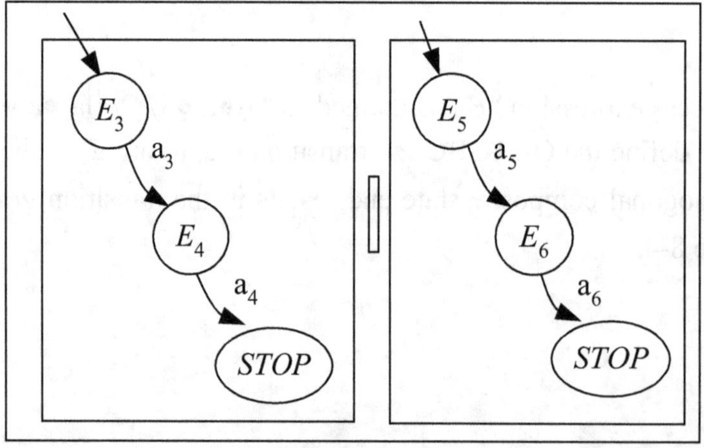

Figure 8-5. Transition Graph TG_{72}

Comparing Figure 8-4 with Figure 8-5, we conclude that the orthogonal composite state does reduce the complexity of the transition graph.

8-5 Transition Graph of FixIFD

In operation-based multi-queue SBC process algebra, the process expression of FixIFD$_i$ is formally defined as "**fix**(X_i=g_{i1}•a_{i2}•a_{i3}•…•a_{in}•X_i)". We use the O-M-SBC-PA transition graph TG_i to define the execution of the process expression FixIFD$_i$, as shown in Figure 8-6.

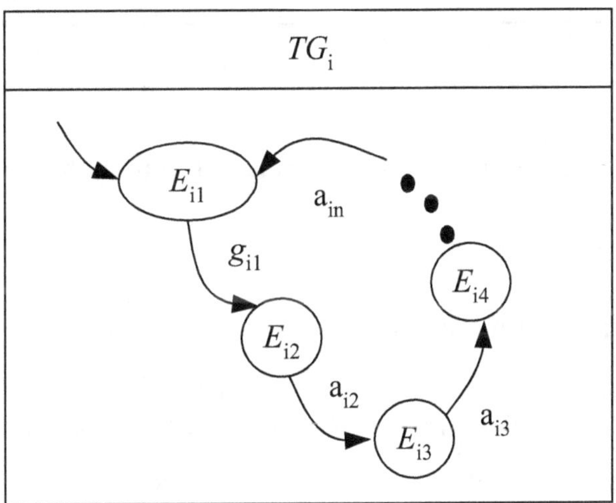

Figure 8-6. Transition Graph for FixIFD$_i$

8-6 TG Relation of FixIFD

We use a transition relation $TGR_i \subseteq \Psi_1 X \Delta X \Psi_2$, where Ψ is a set of "process expressions" and Δ is a relation of "type 1 or 2 interactions" and $(E_{ij}, a, E_{ik}) \in TGR_i$ is denoted by $E_{ij} \xrightarrow{a} E_{ik}$, to represent the O-M-SBC-PA transition graph TG_i of the process expression FixIFD$_i$, as shown in Figure 8-7.

Ψ_1	Δ	Ψ_2
E_{i1}	g_{i1}	E_{i2}
E_{i2}	a_{i2}	E_{i3}
E_{i3}	a_{i3}	E_{i4}
●	●	●
E_{in}	a_{in}	E_{i1}

Figure 8-7. Relation TGR_i for Process Expression "FixIFD_i"

8-7 Transition Graph of a System

In O-M-SBC-PA, the process expression of a system E_{system} is formally defined as "$\mathbf{fix}(X_1=g_{11}\bullet a_{12}\bullet a_{13}\bullet\ldots\bullet a_{1n}\bullet X_1)\square\mathbf{fix}(X_2=g_{21}\bullet a_{22}\bullet a_{23}\bullet\ldots\bullet a_{2n}\bullet X_2)\square\ldots\square\mathbf{fix}(X_m=g_{m1}\bullet a_{m2}\bullet a_{m3}\bullet\ldots\bullet a_{mn}\bullet X_m)$" or "$\text{FixIFD}_1\square\text{FixIFD}_2\square\ldots\square\text{FixIFD}_m$".

By using the O-M-SBC-PA transition graph TG_i to define the execution of the process expression FixIFD_i, we get the O-M-SBC-PA transition graph of a system TG_{system} defined as "$TG_1\square TG_2\square\ldots\square TG_m$" and diagramed as shown in Figure 8-8.

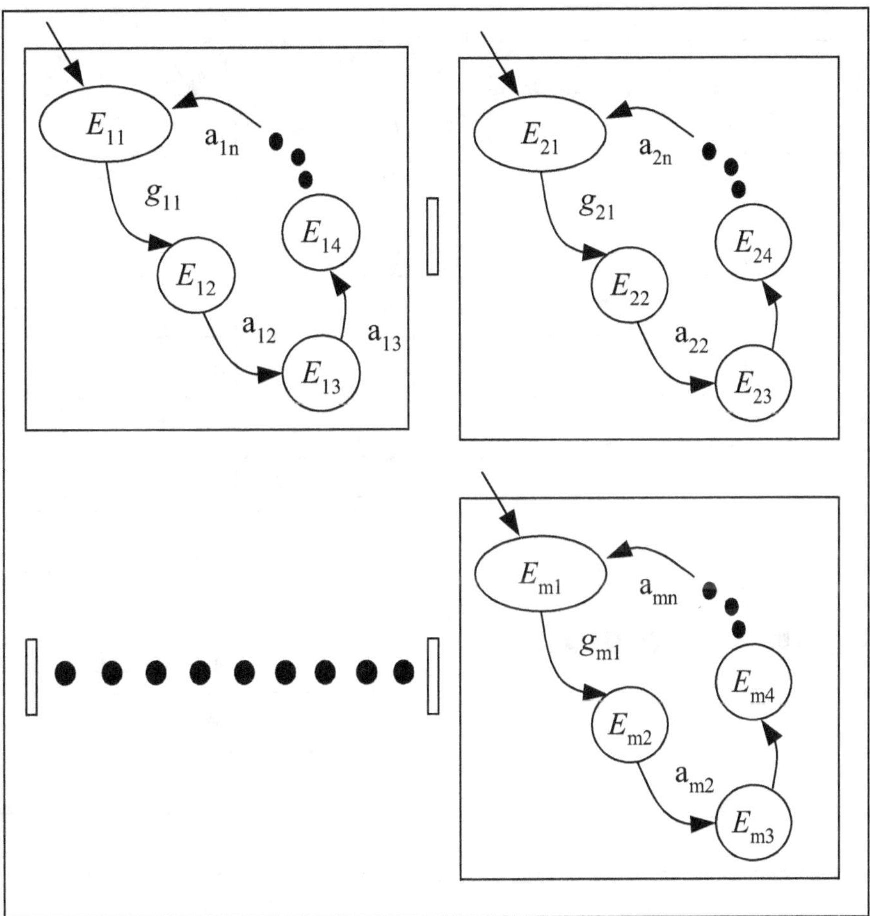

Figure 8-8. Transition Graph TG_{system}

8-8 TG Relation of a System

In O-M-SBC-PA, the transition graph of a system TG_{system} is formally defined as "TG_1□TG_2 □…□TG_m" and each transition graph TG_i is represented by a transition relation $TGR_i \subseteq \Psi_1 X \Delta X \Psi_2$, where Ψ is a set of "process expressions" and Δ is a relation of "type 1 or 2 interactions" and $(E_{ij}, a, E_{ik}) \in TGR_i$ is denoted by $E_{ij} \xrightarrow{a} E_{ik}$. Therefore, we get the O-M-SBC-PA TG relation of a system $TGR_{system} \subseteq \Psi_1 X \Delta X \Psi_2$ defined as "TGR_1□TGR_2□…□TGR_m" and is shown in Figure 8-9.

Ψ_1	Δ	Ψ_2
E_{11}	g_{11}	E_{12}
E_{12}	a_{12}	E_{13}
E_{13}	a_{13}	E_{14}
•	•	•
E_{1n}	a_{1n}	E_{11}

Ψ_1	Δ	Ψ_2
E_{21}	g_{21}	E_{22}
E_{22}	a_{22}	E_{23}
E_{23}	a_{23}	E_{24}
•	•	•
E_{2n}	a_{2n}	E_{21}

• • • • • • • • •

Ψ_1	Δ	Ψ_2
E_{m1}	g_{m1}	E_{m2}
E_{m2}	a_{m2}	E_{m3}
E_{m3}	a_{m3}	E_{m4}
•	•	•
E_{mn}	a_{mn}	E_{m1}

Figure 8-9. Relation TGR_{system}

Chapter 9: Projecting a Use Case Diagram from the O-M-SBC-PA Transition Graph

In this chapter, we discuss how to project a UML use case diagram from the O-M-SBC-PA transition graph of a system.

9-1 UML Use Case Diagrams

The simplest UML use case diagram (UCD) represents the user's interaction with the system, showing the relationship between the user and the different use cases involved. Use case diagrams identify different types of users and different use cases of the system, and are often accompanied by other types of diagrams. Use cases are represented by circles or ellipses.

The UML use case diagram of a system UCD_{system} can be defined by an orthogonal composition "$UCD_1 \square UCD_2 \square \ldots \square UCD_m$", as shown in Figure 9-1.

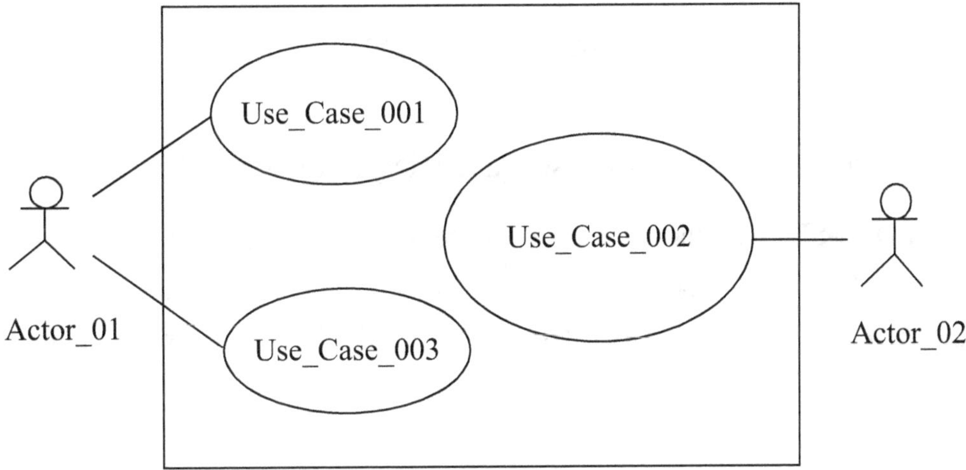

Figure 9-1. UML Use Case Diagram

9-2 UCD Relation (UCDR) of a System

In UML, the use case diagram of a system UCD_{system} can be formally defined as "UCD_1□UCD_2 □…□UCD_m" and each UCD_i is represented by a relation $UCDR_i \subseteq B X U$, where B is a set of "actors" and U is a set of "use cases". Therefore, we get the UCD relation of a system $UCDR_{system} \subseteq B X U$ defined as "$UCDR_1$□$UCDR_2$ □…□$UCDR_m$" and diagramed as shown in Figure 9-2.

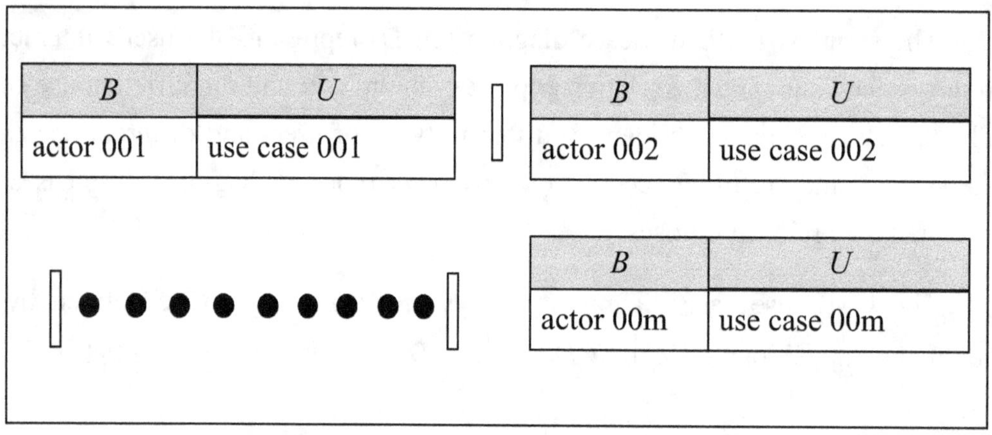

Figure 9-2. Relation $UCDR_{system}$

9-3 Algorithm of Projecting a Use Case Diagram from O-M-SBC-PA

In O-M-SBC-PA, the process of a system is represented by a transition graph TG_{system} (defined as "TG_1□TG_2 □…□ TG_m") with the transition relation $TGR_{system} \subseteq \Psi_1 X \Delta X \Psi_2$ (defined as "TGR_1□TGR_2□…□TGR_m") as shown in Figure 9-3.

Ψ_1	Δ	Ψ_2
E_{11}	g_{11}	E_{12}
E_{12}	a_{12}	E_{13}
E_{13}	a_{13}	E_{14}
•	•	•
E_{1n}	a_{1n}	E_{11}

Ψ_1	Δ	Ψ_2
E_{21}	g_{21}	E_{22}
E_{22}	a_{22}	E_{23}
E_{23}	a_{23}	E_{24}
•	•	•
E_{2n}	a_{2n}	E_{21}

• • • • • • • • •

Ψ_1	Δ	Ψ_2
E_{m1}	g_{m1}	E_{m2}
E_{m2}	a_{m2}	E_{m3}
E_{m3}	a_{m3}	E_{m4}
•	•	•
E_{mn}	a_{mn}	E_{m1}

Figure 9-3. Relation TGR_{system}

We rewrite the TG relation of a system as $TGR_{system} \subseteq \Psi_1 \times N \times \Xi \times \Lambda \times \Theta \times \Gamma \times \Psi_2$ since the "type 1 or 2 interaction" is defined as a relation $\Delta \subseteq N \times \Xi \times L \times \Gamma$ and the "operation call or operation return signature" is defined as a relation $L \subseteq \Lambda \times \Theta$.

Figure 9-4 shows the algorithm of projecting the UCD relation $UCDR_{system} \subseteq BXU$ from the TG relation $TGR_{system} \subseteq \Psi_1 \text{ X } N \text{ X } \Xi \text{ X } \Lambda \text{ X } \Theta \text{ X } \Gamma \text{ X } \Psi_2$.

```
For i = 1, m Loop
  SCANF("%s", @UseCaseName)
  CREATE RELATION UCDR_i (B, U)
  INSERT INTO UCDR_i(B) SELECT Ξ FROM TGR_i fetch first row only;
  UPDATE UCDR_i SET U = @UseCaseName SELECT * FROM UCDR_i
End Loop;

ORTHOGONALLY COMPOSE All UCDR_i (i.e., ||_{i=1, m} UCDR_i) to get UCDR_system
```

Figure 9-4. Algorithm of Projecting the UCD Relation from the TG Relation

Once we have the UCD relation $UCDR_{system}$, it is easy to get a UML use case diagram of the system.

Chapter 10: Projecting a State Diagram from the O-M-SBC-PA Transition Graph

In this chapter, we discuss how to project a UML state diagram from the O-M-SBC-PA transition graph of a system.

10-1 UML State Diagrams

In UML, the state diagram (StD) represents behavior of a system in terms of its transition between states triggered by operation calls. UML state diagram is an object-based variant of Harel statechart, adapted and extended by UML. The goal of UML state diagrams is to overcome the main limitations of traditional finite-state machines while retaining their main benefits. UML state diagrams introduce the new concepts of hierarchically nested states and orthogonal regions, while extending the notion of operation calls.

The UML state diagram of a system StD_{system} can be defined by an orthogonal composite state "StD_1□StD_2 □…□StD_m", as shown in Figure 10-1.

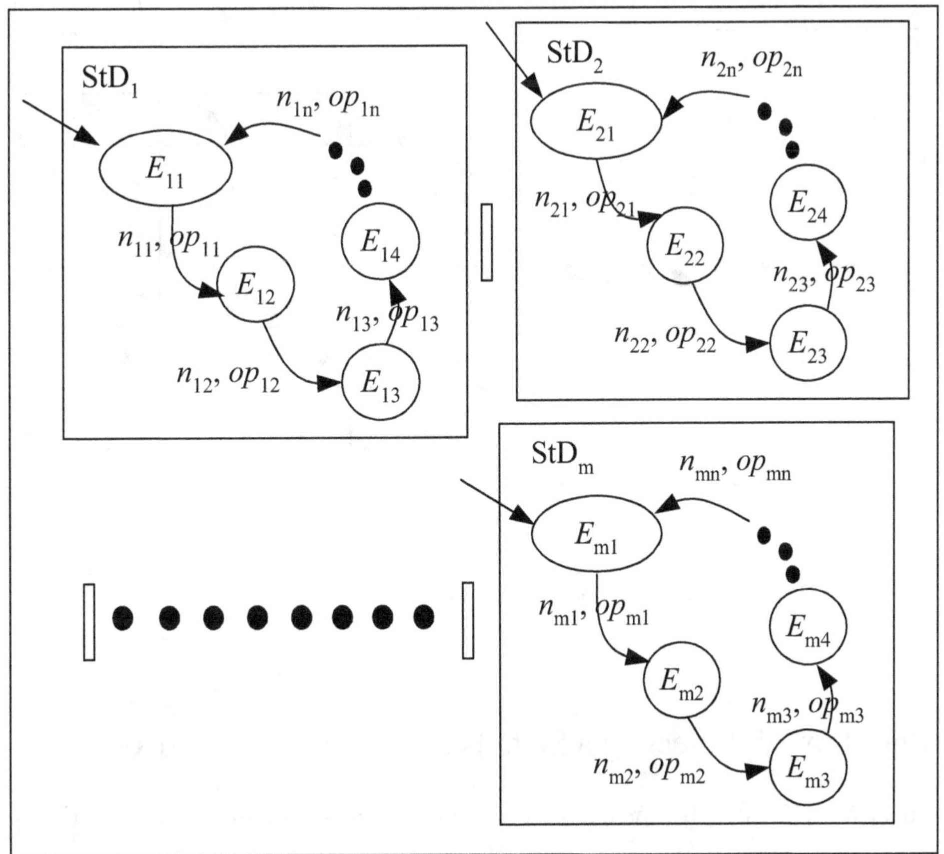

Figure 10-1. UML State Diagram StD_{system}

10-2 StD Relation (StDR) of a System

In UML, the state diagram of a system StD_{system} is formally defined as "StD_1□StD_2 □…□StD_m" and each substate diagram StD_i is represented by a relation $StDR_i \subseteq \Psi_1 X N X \Lambda X \Psi_2$, where Ψ is a set of "process expressions" and N is a set of "operation call or operation return tags" and Λ is a set of "operation names" and $(E_{ij}, n, op, E_{ik}) \in StDR_i$ is denoted by $E_{ij} \xrightarrow{n,\ op} E_{ik}$. Therefore, we get the StD relation of a system $StDR_{system} \subseteq \Psi_1 \ X \ N \ X \Lambda X \ \Psi_2$ defined as "$StDR_1$□$StDR_2$□…□$StDR_m$" and diagramed as shown in Figure 10-2.

Ψ_1	N	Λ	Ψ_2
E_{11}	n_{11}	op_{11}	E_{12}
E_{12}	n_{12}	op_{12}	E_{13}
E_{13}	n_{13}	op_{13}	E_{14}
•	•	•	•
E_{1n}	n_{1n}	op_{1n}	E_{11}

□

Ψ_1	N	Λ	Ψ_2
E_{21}	n_{21}	op_{21}	E_{22}
E_{22}	n_{22}	op_{22}	E_{23}
E_{23}	n_{23}	op_{23}	E_{24}
•	•	•	•
E_{2n}	n_{2n}	op_{2n}	E_{21}

□ • • • • • • • • • □

Ψ_1	N	Λ	Ψ_2
E_{m1}	n_{m1}	op_{m1}	E_{m2}
E_{m2}	n_{m2}	op_{m2}	E_{m3}
E_{m3}	n_{m3}	op_{m3}	E_{m4}
•	•	•	•
E_{mn}	n_{mn}	op_{mn}	E_{m1}

Figure 10-2. Relation $StDR_{system}$

10-3 Algorithm of Projecting a State Diagram from O-M-SBC-PA

In O-M-SBC-PA, the process of a system is represented by a transition graph TG_{system} (defined as "TG_1□TG_2 □…□ TG_m") with the transition relation $TGR_{system} \subseteq \Psi_1 X \Lambda X \Psi_2$ (defined as "TGR_1□TGR_2□…□TGR_m") as shown in Figure 10-3.

Ψ_1	Δ	Ψ_2
E_{11}	g_{11}	E_{12}
E_{12}	a_{12}	E_{13}
E_{13}	a_{13}	E_{14}
•	•	•
E_{1n}	a_{1n}	E_{11}

Ψ_1	Δ	Ψ_2
E_{21}	g_{21}	E_{22}
E_{22}	a_{22}	E_{23}
E_{23}	a_{23}	E_{24}
•	•	•
E_{2n}	a_{2n}	E_{21}

• • • • • • • • •

Ψ_1	Δ	Ψ_2
E_{m1}	g_{m1}	E_{m2}
E_{m2}	a_{m2}	E_{m3}
E_{m3}	a_{m3}	E_{m4}
•	•	•
E_{mn}	a_{mn}	E_{m1}

Figure 10-3. Relation TGR_{system}

We rewrite the TG relation of a system as $TGR_{system} \subseteq \Psi_1 \text{ X } N \text{ X } \Xi \text{ X } \Lambda \text{ X } \Theta \text{ X } \Gamma \text{ X } \Psi_2$ since the "type 1 or 2 interaction" is defined as a relation $\Delta \subseteq N \text{ X } \Xi \text{ X } L \text{ X } \Gamma$ and the "operation call or operation return signature" is defined as a relation $L \subseteq \Lambda \text{ X } \Theta$.

Figure 10-4 shows the algorithm of projecting the StD relation $StDR_{system} \subseteq \Psi_1 X N X \Lambda X \Psi_2$ from the TG relation $TGR_{system} \subseteq \Psi_1 X N X \Xi X \Lambda X \Theta X \Gamma X \Psi_2$.

For i = 1, m **Loop**
SELECT Ψ_1, N, Λ, Ψ_2 INTO $StDR_i$ FROM TGR_i;
End Loop;

ORTHOGONALLY COMPOSE All $StDR_i$ (i.e., $\|_{i=1,m} StDR_i$) to get $StDR_{system}$

Figure 10-4. Algorithm of Projecting the StD Relation from the TG Relation

Once we have the StD relation $StDR_{system}$, it is easy to get a UML state diagram of the system.

Chapter 11: Projecting an Activity Diagram from the O-M-SBC-PA Transition Graph

In this chapter, we discuss how to project a UML activity diagram from the O-M-SBC-PA transition graph of a system.

11-1 UML Activity Diagrams

In UML, activity diagrams (AD) are intended to model both computational and organizational processes, as well as the data flows intersecting with the related activities. Although activity diagrams primarily show the overall flow of control, they can also include elements showing the flow of data between activities through one or more data stores.

The UML activity diagram of a system AD_{system} can be defined by an orthogonal composition "AD_1□AD_2□…□AD_m", as shown in Figure 11-1.

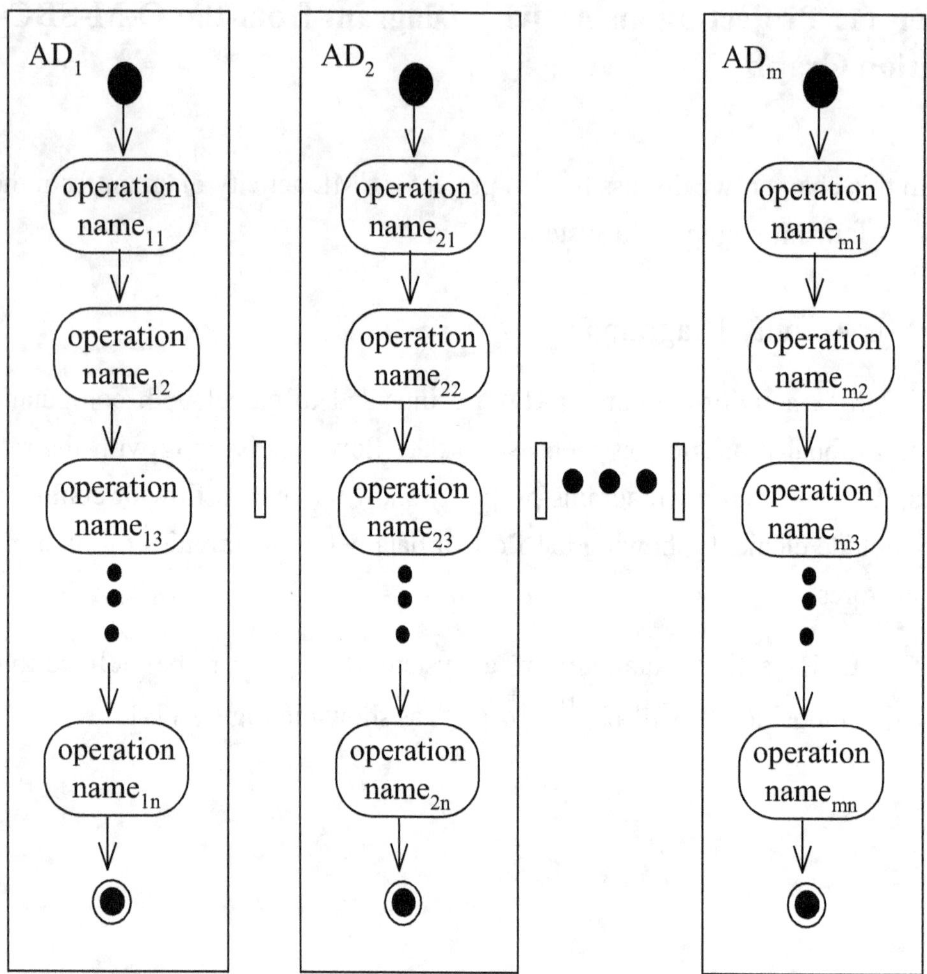

Figure 11-1. UML Activity Diagram AD_{system}

11-2 AD Relation (ADR) of a System

In UML, the activity diagram of a system AD_{system} can be formally defined as "$AD_1 \square AD_2\ \square \ldots \square AD_m$" and each AD_i is represented by a relation $ADR_i \subseteq \Psi_1 X N X \Lambda X \Psi_2$, where Ψ is a set of "process expressions" and N is a set of "operation call or operation return tags" and Λ is a set of "operation names" and $(E_{ij}, n, op, E_{ik}) \in ADR_i$ is denoted by $E_{ij} \xrightarrow{n,\ op} E_{ik}$. Therefore, we get the AD relation of a system $ADR_{system} \subseteq \Psi_1 X N X \Lambda X \Psi_2$ defined as "$ADR_1 \square ADR_2 \square \ldots \square ADR_m$" and diagramed as shown in Figure 11-2.

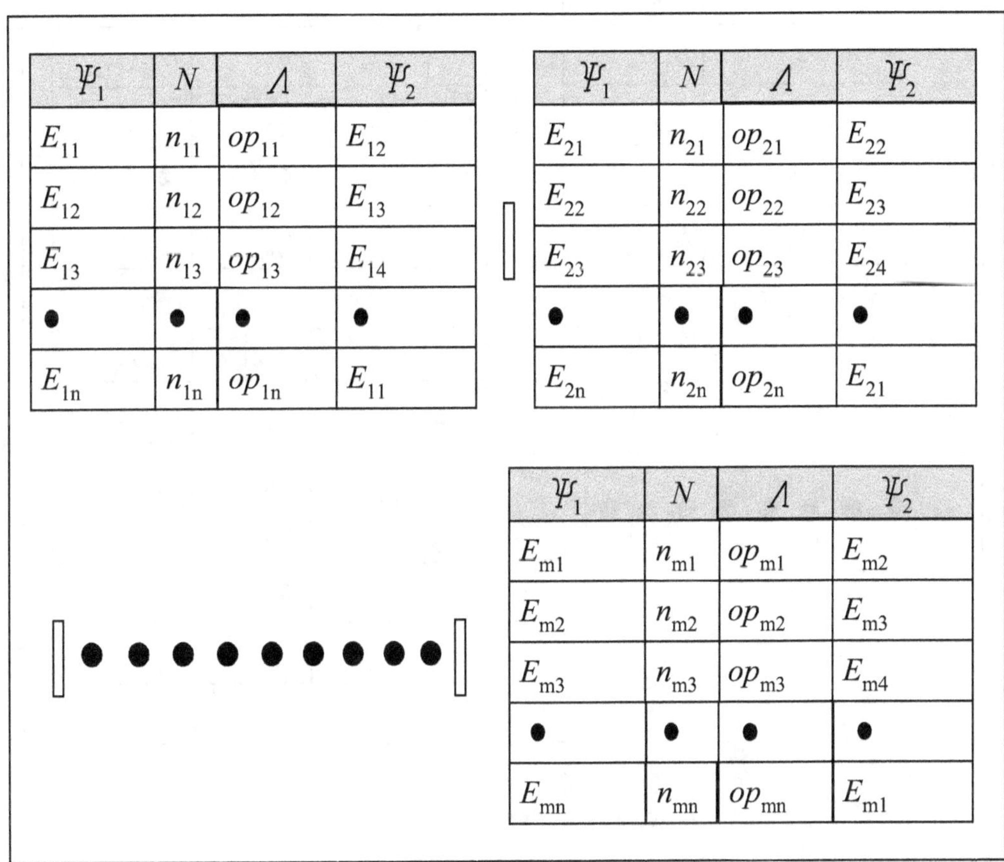

Ψ_1	N	Λ	Ψ_2
E_{11}	n_{11}	op_{11}	E_{12}
E_{12}	n_{12}	op_{12}	E_{13}
E_{13}	n_{13}	op_{13}	E_{14}
•	•	•	•
E_{1n}	n_{1n}	op_{1n}	E_{11}

Ψ_1	N	Λ	Ψ_2
E_{21}	n_{21}	op_{21}	E_{22}
E_{22}	n_{22}	op_{22}	E_{23}
E_{23}	n_{23}	op_{23}	E_{24}
•	•	•	•
E_{2n}	n_{2n}	op_{2n}	E_{21}

• • • • • • • • •

Ψ_1	N	Λ	Ψ_2
E_{m1}	n_{m1}	op_{m1}	E_{m2}
E_{m2}	n_{m2}	op_{m2}	E_{m3}
E_{m3}	n_{m3}	op_{m3}	E_{m4}
•	•	•	•
E_{mn}	n_{mn}	op_{mn}	E_{m1}

Figure 11-2. Relation ADR_{system}

11-3 Algorithm of Projecting an Activity Diagram from O-M-SBC-PA

In O-M-SBC-PA, the process of a system is represented by a transition graph TG_{system} (defined as "$TG_1 \square TG_2 \square \ldots \square TG_m$") with the transition relation $TGR_{\text{system}} \subseteq \Psi_1 X \Lambda X \Psi_2$ (defined as "$TGR_1 \square TGR_2 \square \ldots \square TGR_m$") as shown in Figure 11-3.

Ψ_1	Δ	Ψ_2
E_{11}	g_{11}	E_{12}
E_{12}	a_{12}	E_{13}
E_{13}	a_{13}	E_{14}
•	•	•
E_{1n}	a_{1n}	E_{11}

Ψ_1	Δ	Ψ_2
E_{21}	g_{21}	E_{22}
E_{22}	a_{22}	E_{23}
E_{23}	a_{23}	E_{24}
•	•	•
E_{2n}	a_{2n}	E_{21}

• • • • • • • • •

Ψ_1	Δ	Ψ_2
E_{m1}	g_{m1}	E_{m2}
E_{m2}	a_{m2}	E_{m3}
E_{m3}	a_{m3}	E_{m4}
•	•	•
E_{mn}	a_{mn}	E_{m1}

Figure 11-3. Relation TGR_{system}

We rewrite the TG relation of a system as $TGR_{\text{system}} \subseteq \Psi_1 \text{ X } N \text{ X } \Xi \text{ X } \Lambda \text{ X } \Theta \text{ X } \Gamma \text{ X } \Psi_2$ since the "type 1 or 2 interaction" is defined as a relation $\Delta \subseteq N \text{ X } \Xi \text{ X } L \text{ X } \Gamma$ and the "operation call or operation return signature" is defined as a relation $L \subseteq \Lambda \text{ X } \Theta$.

Figure 11-4 shows the algorithm of projecting the AD relation $ADR_{system} \subseteq \Psi_1 X N X \Lambda X \Psi_2$ from the TG relation $TGR_{system} \subseteq \Psi_1 X N X \Xi X \Lambda X \Theta X \Gamma X \Psi_2$.

For i = 1, m **Loop**
SELECT Ψ_1 , N , Λ , Ψ_2 INTO ADR_i FROM TGR_i;
End Loop;

ORTHOGONALLY COMPOSE All ADR_i (i.e., $\|_{i=1,m}$ ADR_i) to get ADR_{system}

Figure 11-4. Algorithm of Projecting the AD Relation from the TG Relation

Once we have the AD relation ADR_{system}, it is easy to get a UML activity diagram of the system.

Chapter 12: Projecting a Sequence Diagram from the O-M-SBC-PA Transition Graph

In this chapter, we discuss how to project a UML sequence diagram from the O-M-SBC-PA transition graph of a system.

12-1 UML Sequence Diagrams

UML sequence diagrams (SqD) are interaction diagrams that detail how to perform operations. They capture interactions between objects in a collaborative environment. Sequence diagrams are time focuses that visually display the order of interactions by using the vertical axis of the diagram to indicate when and when the message was sent. Sequence diagrams are sometimes called event diagrams or event scenarios.

The UML sequence diagram of a system SqD_{system} can be defined by an orthogonal composition "SqD_1□SqD_2 □…□SqD_{m}", as shown in Figure 12-1.

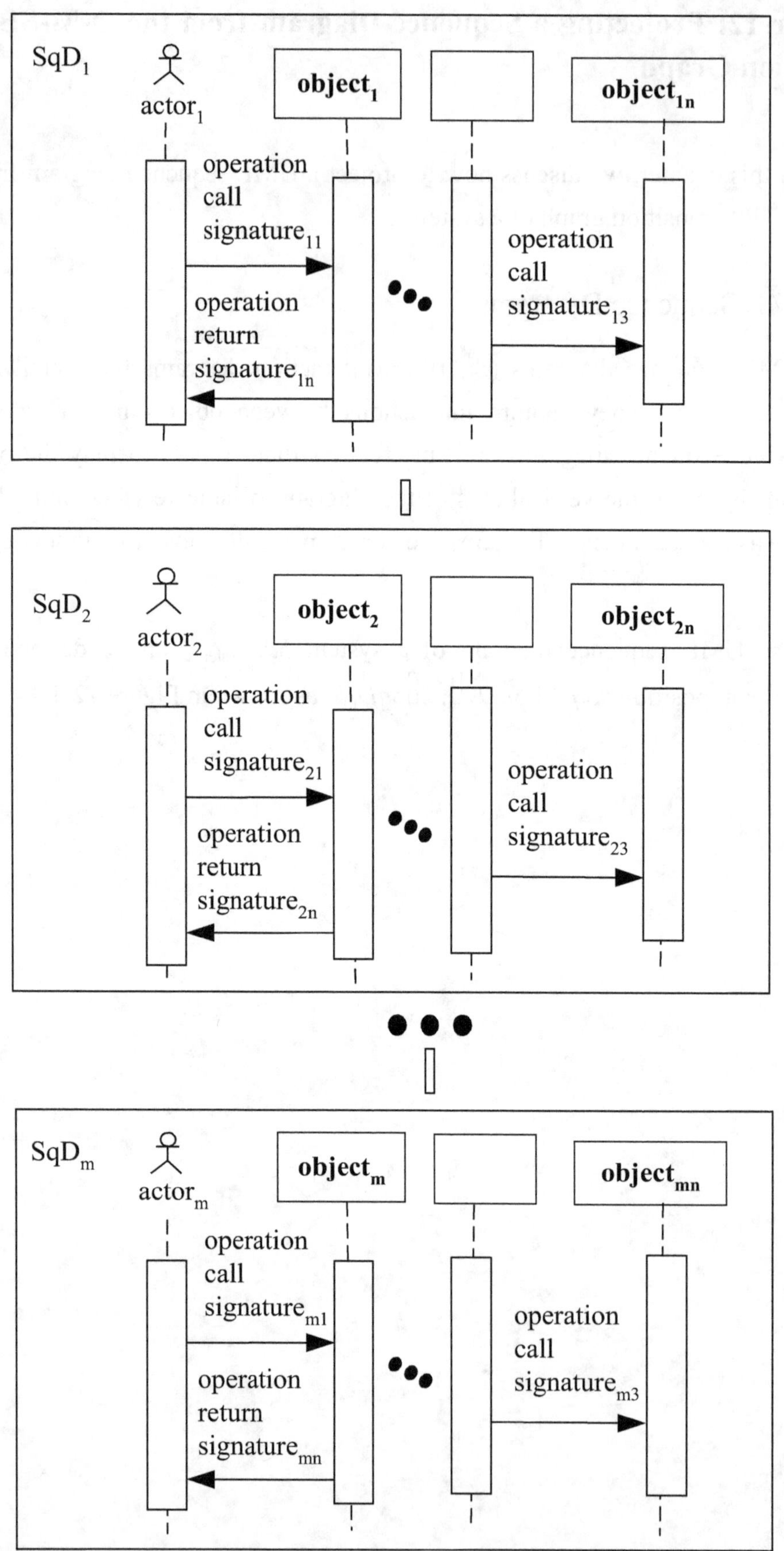

Figure 12-1. UML Sequence Diagram SqD_{system}

12-2 SqD Relation (SqDR) of a System

In UML, the sequence diagram of a system SqD_{system} is formally defined as "SqD_1▯SqD_2▯…▯SqD_m" and each SqD_i is represented by a relation $SqDR_i \subseteq E X N X \Xi X \mathit{\Lambda} X \mathit{\Theta} X \mathit{\Gamma}$, where E is a set of "execution orders" and N is a set of "operation call or operation return tags" and Ξ is a set of "actors or objects" and $\mathit{\Lambda}$ is a set of "operation names" and $\mathit{\Theta}$ is a set of "parameter lists" and $\mathit{\Gamma}$ is a set of "objects". Therefore, we get the SqD relation of a system $SqDR_{system} \subseteq E X N X \Xi X \mathit{\Lambda} X \mathit{\Theta} X \mathit{\Gamma}$ defined as "$SqDR_1$▯$SqDR_2$▯…▯$SqDR_m$" and diagramed as shown in Figure 12-2.

E	N	Ξ	Λ	Θ	Γ
1	n_{11}	ρ_{11}	op_{11}	p_{11}	o_{11}
2	n_{12}	ρ_{12}	op_{12}	p_{12}	o_{12}
•	•	•	•	•	•
n	n_{1n}	ρ_{1n}	op_{1n}	p_{1n}	o_{1n}

E	N	Ξ	Λ	Θ	Γ
1	n_{21}	ρ_{21}	op_{21}	p_{21}	o_{21}
2	n_{22}	ρ_{22}	op_{22}	p_{22}	o_{22}
•	•	•	•	•	•
n	n_{2n}	ρ_{2n}	op_{2n}	p_{2n}	o_{2n}

• • •

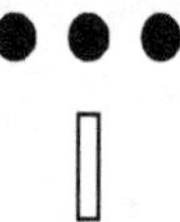

E	N	Ξ	Λ	Θ	Γ
1	n_{m1}	ρ_{m1}	op_{m1}	p_{m1}	o_{m1}
2	n_{m2}	ρ_{m2}	op_{m2}	p_{m2}	o_{m2}
•	•	•	•	•	•
n	n_{mn}	ρ_{mn}	op_{mn}	p_{mn}	o_{mn}

Figure 12-2. Relation $SqDR_{\text{system}}$

12-3 Algorithm of Projecting a Sequence Diagram from O-M-SBC-PA

In O-M-SBC-PA, the process of a system is represented by a transition graph TG_{system} (defined as "$TG_1 \square TG_2 \ \square \ldots \square \ TG_m$") with the transition relation $TGR_{system} \subseteq \Psi_1 X \Delta X \Psi_2$ (defined as "$TGR_1 \square TGR_2 \square \ldots \square TGR_m$") as shown in Figure 12-3.

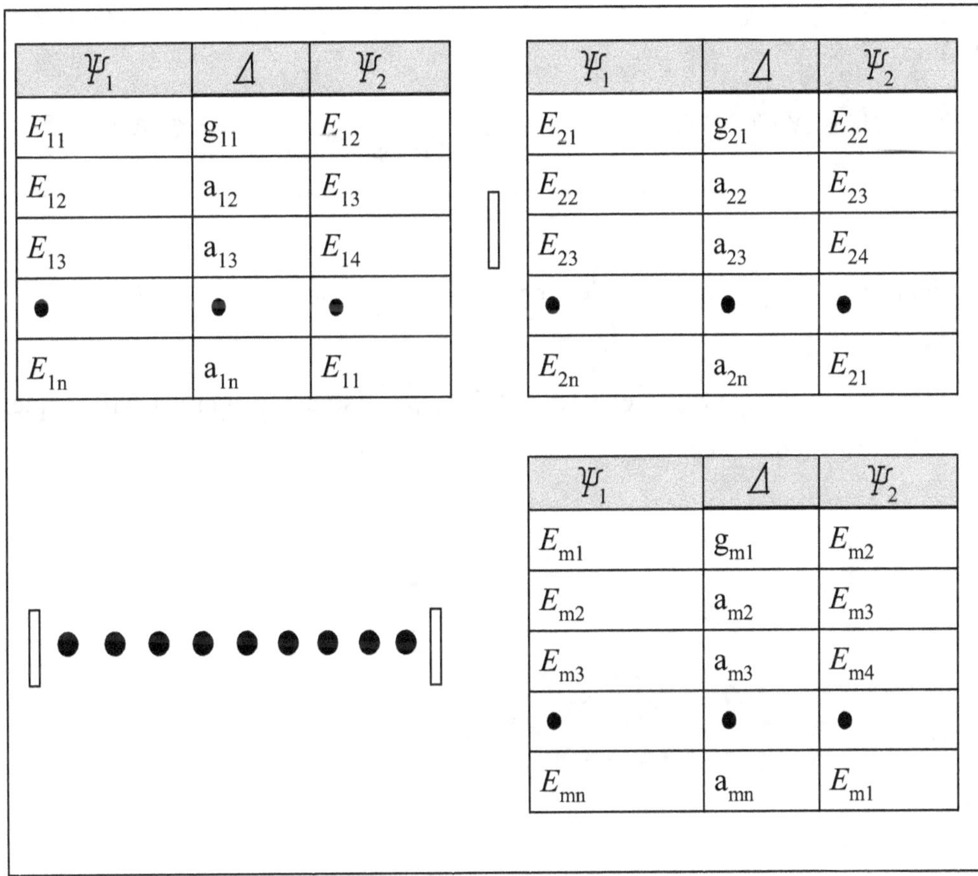

Ψ_1	Δ	Ψ_2
E_{11}	g_{11}	E_{12}
E_{12}	a_{12}	E_{13}
E_{13}	a_{13}	E_{14}
•	•	•
E_{1n}	a_{1n}	E_{11}

$\square$

Ψ_1	Δ	Ψ_2
E_{21}	g_{21}	E_{22}
E_{22}	a_{22}	E_{23}
E_{23}	a_{23}	E_{24}
•	•	•
E_{2n}	a_{2n}	E_{21}

$\square$ • • • • • • • • • $\square$

Ψ_1	Δ	Ψ_2
E_{m1}	g_{m1}	E_{m2}
E_{m2}	a_{m2}	E_{m3}
E_{m3}	a_{m3}	E_{m4}
•	•	•
E_{mn}	a_{mn}	E_{m1}

Figure 12-3. Relation TGR_{system}

We rewrite the TG relation of a system as $TGR_{system} \subseteq \Psi_1 \text{ X } N \text{ X } \Xi \text{ X } \Lambda \text{ X } \Theta \text{ X } \Gamma \text{ X } \Psi_2$ since the "type 1 or 2 interaction" is defined as a relation $\Delta \subseteq N \text{ X } \Xi \text{ X } L \text{ X } \Gamma$ and the "operation call or operation return signature" is defined as a relation $L \subseteq \Lambda \text{ X } \Theta$.

Figure 12-4 shows the algorithm of projecting the SqD relation $SqDR_{system} \subseteq E X N X \Xi X \Lambda X \Theta X \Gamma$ from the TG relation $TGR_{system} \subseteq \Psi_1 X N X \Xi X \Lambda X \Theta X \Gamma X \Psi_2$.

For i = 1, m **Loop**
CREATE RELATION $SqDR_i$ (E int IDENTITY(1,1), N, Ξ, Λ, Θ, Γ);
INSERT INTO $SqDR_i$ (N, Ξ, Λ, Θ, Γ) SELECT N, Ξ, Λ, Θ, Γ FROM TGR_i;
End Loop;

ORTHOGONALLY COMPOSE All $SqDR_i$ (i.e., $\|_{i=1,m} SqDR_i$) to get $SqDR_{system}$

Figure 12-4. Algorithm of Projecting the SqD Relation from the TG Relation

Once we have the SqD relation $SqDR_{system}$, it is easy to get a UML sequence diagram of the system.

Chapter 13: Projecting a Communication Diagram from the O-M-SBC-PA Transition Graph

In this chapter, we discuss how to project a UML communication diagram from the O-M-SBC-PA transition graph of a system.

13-1 UML Communication Diagrams

In UML, the communication diagram (ComD) shows a lot of information that is identical to the sequence diagram, but because of how the information is presented, some of the information is easier to find in one diagram than in the other. The communication diagram shows which object interacts better with each object, but the sequence diagram shows that the order in which the interactions occur is clearer.

The UML communication diagram of a system $ComD_{system}$ can be defined by an orthogonal composition "$ComD_1 \square ComD_2 \square \ldots \square ComD_m$", as shown in Figure 13-1.

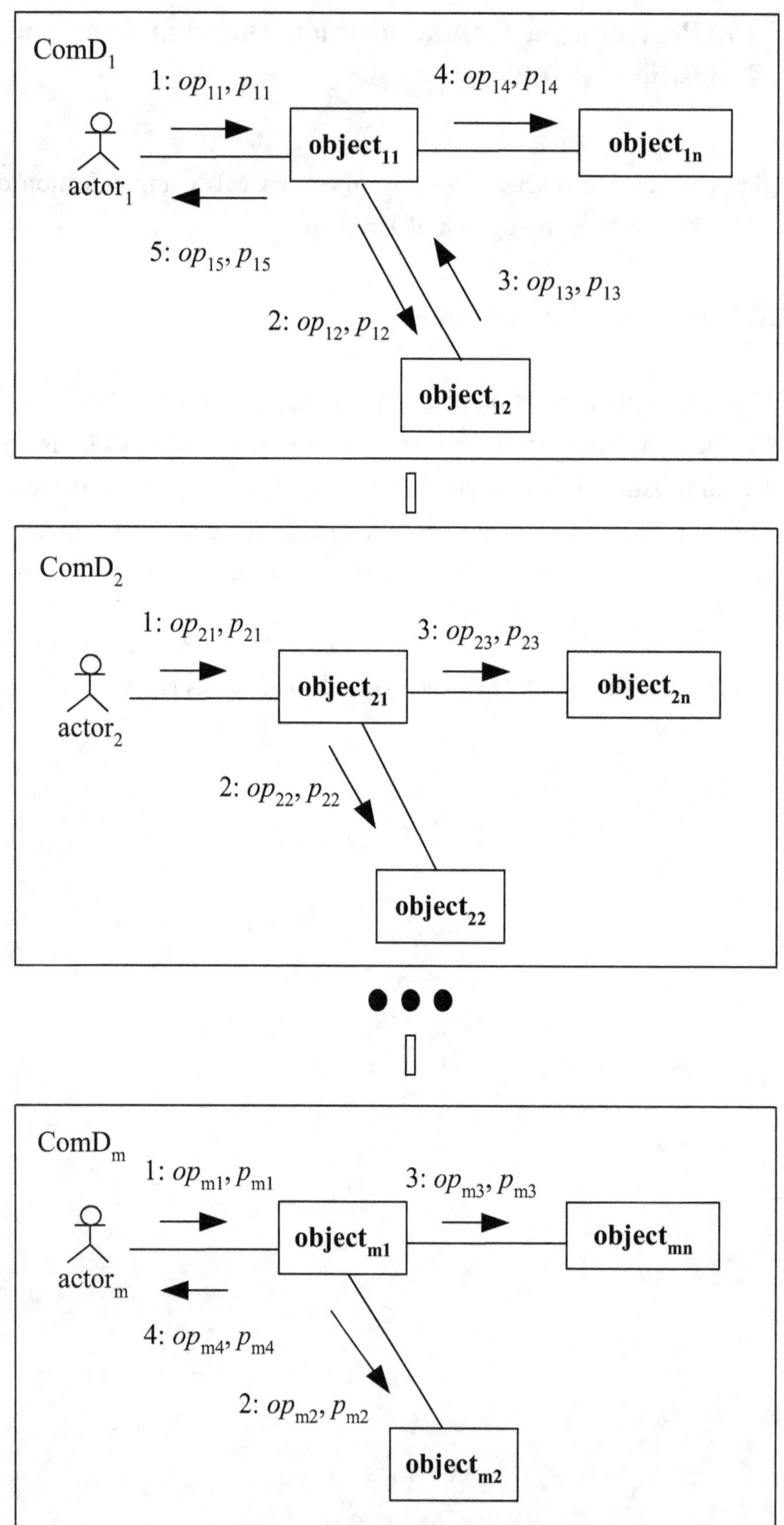

Figure 13-1. UML Communication Diagram $ComD_{system}$

13-2 ComD Relation (ComDR) of a System

In UML, the communication diagram of a system $ComD_{system}$ is formally defined as "$ComD_1$▯$ComD_2$ ▯...▯$ComD_m$" and each $ComD_i$ is represented by a relation $ComDR_i \subseteq EXNX\Xi X\Lambda X\Theta X\Gamma$, where E is a set of "execution orders" and N is a set of "operation call or operation return tags" and Ξ is a set of "actors or objects" and Λ is a set of "operation names" and Θ is a set of "parameter lists" and Γ is a set of "objects". Therefore, we get the ComD relation of a system $ComDR_{system} \subseteq EXNX\Xi X\Lambda X\Theta X\Gamma$ defined as "$ComDR_1$▯$ComDR_2$▯...▯$ComDR_m$" and diagramed as shown in Figure 13-2.

E	N	Ξ	Λ	Θ	Γ
1	n_{11}	ρ_{11}	op_{11}	p_{11}	o_{11}
2	n_{12}	ρ_{12}	op_{12}	p_{12}	o_{12}
•	•	•	•	•	•
n	n_{1n}	ρ_{1n}	op_{1n}	p_{1n}	o_{1n}

▯

E	N	Ξ	Λ	Θ	Γ
1	n_{21}	ρ_{21}	op_{21}	p_{21}	o_{21}
2	n_{22}	ρ_{22}	op_{22}	p_{22}	o_{22}
•	•	•	•	•	•
n	n_{2n}	ρ_{2n}	op_{2n}	p_{2n}	o_{2n}

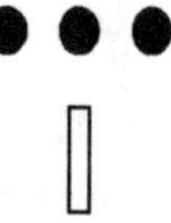

E	N	Ξ	Λ	Θ	Γ
1	n_{m1}	ρ_{m1}	op_{m1}	p_{m1}	o_{m1}
2	n_{m2}	ρ_{m2}	op_{m2}	p_{m2}	o_{m2}
•	•	•	•	•	•
n	n_{mn}	ρ_{mn}	op_{mn}	p_{mn}	o_{mn}

Figure 13-2. Relation $ComDR_{system}$

13-3 Algorithm of Projecting a Communication Diagram from O-M-SBC-PA

In O-M-SBC-PA, the process of a system is represented by a transition graph TG_{system} (defined as "TG_1□TG_2□...□TG_m") with the transition relation $TGR_{\text{system}} \subseteq \Psi_1 X \Delta X \Psi_2$ (defined as "TGR_1□TGR_2□...□TGR_m") as shown in Figure 13-3.

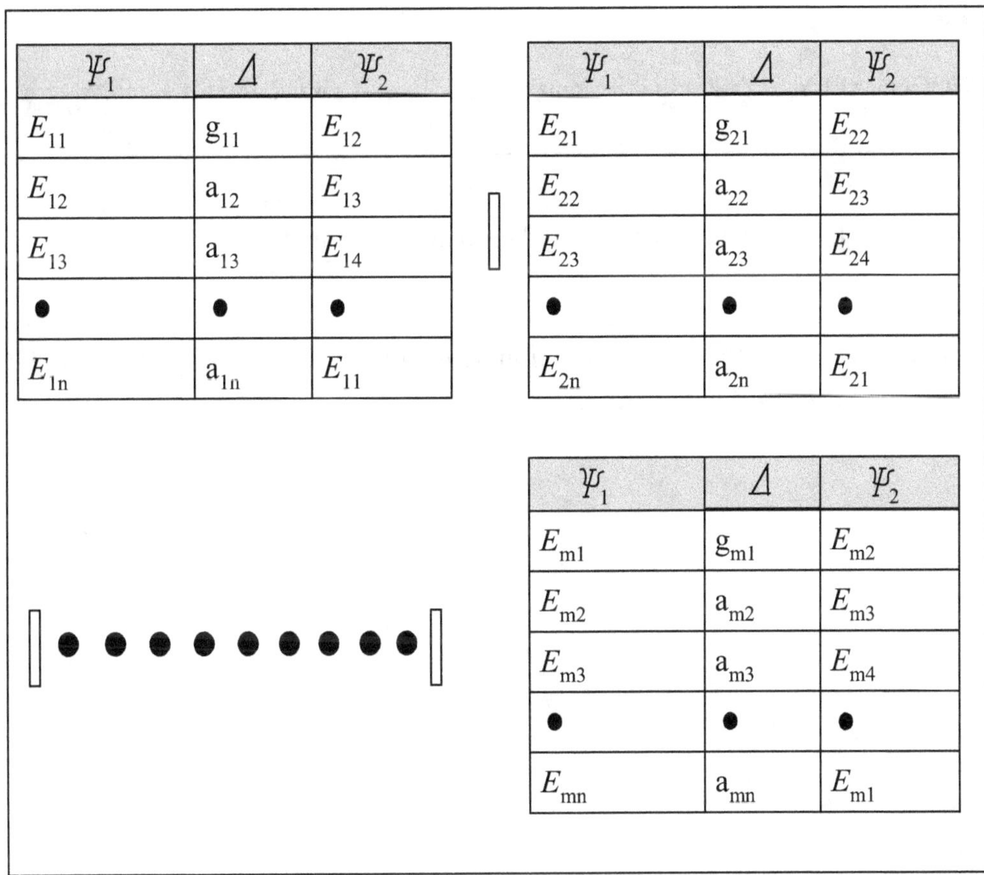

Ψ_1	Δ	Ψ_2
E_{11}	g_{11}	E_{12}
E_{12}	a_{12}	E_{13}
E_{13}	a_{13}	E_{14}
•	•	•
E_{1n}	a_{1n}	E_{11}

□

Ψ_1	Δ	Ψ_2
E_{21}	g_{21}	E_{22}
E_{22}	a_{22}	E_{23}
E_{23}	a_{23}	E_{24}
•	•	•
E_{2n}	a_{2n}	E_{21}

□ • • • • • • • • □

Ψ_1	Δ	Ψ_2
E_{m1}	g_{m1}	E_{m2}
E_{m2}	a_{m2}	E_{m3}
E_{m3}	a_{m3}	E_{m4}
•	•	•
E_{mn}	a_{mn}	E_{m1}

Figure 13-3. Relation TGR_{system}

We rewrite the TG relation of a system as $TGR_{\text{system}} \subseteq \Psi_1 X N X \Xi X \Lambda X \Theta X \Gamma X \Psi_2$ since the "type 1 or 2 interaction" is defined as a relation $\Delta \subseteq N X \Xi X L X \Gamma$ and the "operation call or operation return signature" is defined as a relation $L \subseteq \Lambda X \Theta$.

Figure 13-4 shows the algorithm of projecting the ComD relation $ComDR_{system} \subseteq E X N X \Xi X \Lambda X \Theta X \Gamma$ from the TG relation $TGR_{system} \subseteq \Psi_1 X N X \Xi X \Lambda X \Theta X \Gamma X \Psi_2$.

For i = 1, m **Loop**
CREATE RELATION $ComDR_i$ (E int IDENTITY(1,1), N, Ξ, Λ, Θ, Γ);
INSERT INTO $ComDR_i$ (N, Ξ, Λ, Θ, Γ) SELECT N, Ξ, Λ, Θ, Γ FROM TGR_i;
End Loop;

ORTHOGONALLY COMPOSE All $ComDR_i$ (i.e., $\|_{i=1,m} ComDR_i$) to get $ComDR_{system}$

Figure 13-4. Algorithm of Projecting the ComD Relation from the TG Relation

Once we have the ComD relation $ComDR_{system}$, it is easy to get a UML communication diagram of the system.

Chapter 14: Projecting a Class Diagram from the O-M-SBC-PA Transition Graph

In this chapter, we discuss how to project a UML class diagram from the O-M-SBC-PA transition graph of a system.

14-1 UML Class Diagrams

A class diagram (ClsD) in Unified Modeling Language (UML) is a static structure diagram that describes the structure of a system by showing the classes of the system, their attributes, operations (or methods), and relationships between objects, as shown in Figure 14-1.

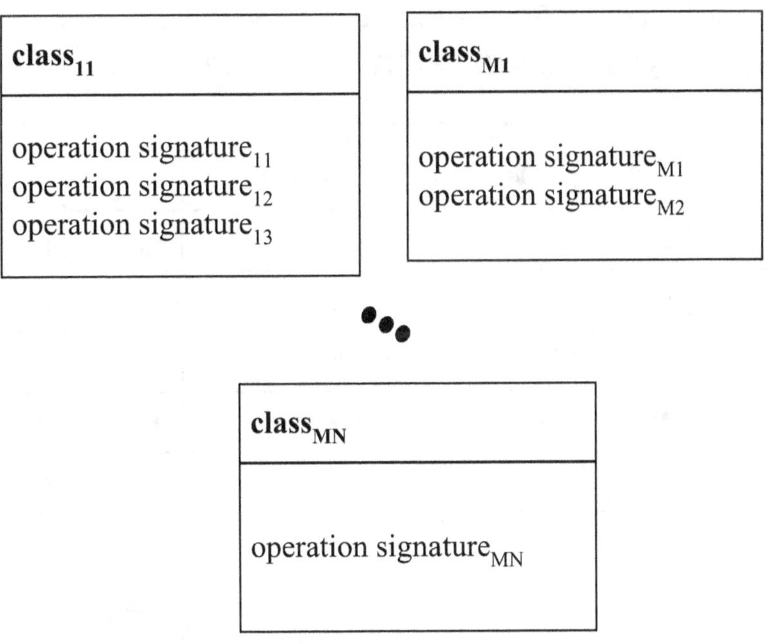

Figure 14-1. UML Class Diagram $ClsD_{system}$

14-2 ClsD Relation (ClsDR) of a System

In UML, the class diagram of a system $ClsD_{system}$ is formally represented by a relation $ClsDR_{system} \subseteq C X \Lambda X \Theta$, where C is a set of "classes" and Λ is a set of "operation names" and Θ is a set of "parameter lists", as shown in Figure 14-2.

C	Λ	Θ
c_{001}	op_{001}	p_{001}
c_{001}	op_{002}	p_{002}
c_{002}	op_{003}	p_{003}
c_{003}	op_{004}	p_{004}

Figure 14-2. Relation $ClsDR_{\text{system}}$

14-3 Algorithm of Projecting Class Diagrams from O-M-SBC-PA

In O-M-SBC-PA, the process of a system is represented by a transition graph TG_{system} (defined as "TG_1□TG_2□…□TG_m") with the transition relation $TGR_{\text{system}} \subseteq \Psi_1 X \Delta X \Psi_2$ (defined as "TGR_1□TGR_2□…□TGR_m") as shown in Figure 14-3.

Ψ_1	Δ	Ψ_2
E_{11}	g_{11}	E_{12}
E_{12}	a_{12}	E_{13}
E_{13}	a_{13}	E_{14}
•	•	•
E_{1n}	a_{1n}	E_{11}

□

Ψ_1	Δ	Ψ_2
E_{21}	g_{21}	E_{22}
E_{22}	a_{22}	E_{23}
E_{23}	a_{23}	E_{24}
•	•	•
E_{2n}	a_{2n}	E_{21}

□ • • • • • • • • • □

Ψ_1	Δ	Ψ_2
E_{m1}	g_{m1}	E_{m2}
E_{m2}	a_{m2}	E_{m3}
E_{m3}	a_{m3}	E_{m4}
•	•	•
E_{mn}	a_{mn}	E_{m1}

Figure 14-3. Relation TGR_{system}

We rewrite the TG relation of a system as $TGR_{\text{system}} \subseteq \Psi_1 \text{ X } N \text{ X } \Xi \text{ X } \Lambda \text{ X } \Theta \text{ X } \Gamma \text{ X } \Psi_2$ since the "type 1 or 2 interaction" is defined as a relation $\Delta \subseteq N \text{ X } \Xi \text{ X } L \text{ X } \Gamma$ and the "operation call or operation return signature" is defined as a relation $L \subseteq \Lambda \text{ X } \Theta$.

Figure 14-4 shows the algorithm of projecting the ClsD relation $ClsDR_{\text{system}} \subseteq C \text{ X} \Lambda \text{X} \Theta$ from the TG relation $TGR_{\text{system}} \subseteq \Psi_1 \text{X} N \text{X} \Xi \text{X} \Lambda \text{X} \Theta \text{ X} \Gamma \text{X} \Psi_2$.

```
For i = 1, m  Loop
  SELECT DISTINCT Γ, Λ, Θ  INTO ClsDR_i (C, Λ, Θ)
  FROM TGR_i WHERE  N = 'OPERATION_CALL';

  SELECT DISTINCT Γ, Λ, Θ  INTO ClsDR_RETURN_i (C, Λ, Θ)
  FROM TGR_i WHERE  N = 'OPERATION_RETURN';

  UPDATE ClsDR_i
  SET L = MERGE (Operation Call Formula, Operation Return Formula)
  WHERE  there exists corresponding
  Operation Return Formula in ClsDR_RETURN_i ;

  INSERT INTO ClsDR_1~m (C, Λ, Θ) SELECT * FROM ClsDR_i;
End Loop;

SELECT DISTINCT * INTO ClsDR_system FROM ClsDR_1~m;
```

Figure 14-4. Algorithm of Projecting the ClsD Relation from the TG Relation

Once we have the ClsD relation $ClsDR_{\text{system}}$, it is easy to get a UML class diagram of the system.

Chapter 15: Projecting an Object Diagram from the O-M-SBC-PA Transition Graph

In this chapter, we discuss how to project a UML object diagram from the O-M-SBC-PA transition graph of a system.

15-1 UML Object Diagrams

An object diagram (OD), as shown in Figure 15-1, is a graph of instances, including objects and data values. A static object diagram is an instance of a class diagram; it shows a snapshot of the system's detailed state at a certain point in time. The use of object diagrams is fairly limited, that is, an example of a data structure is displayed.

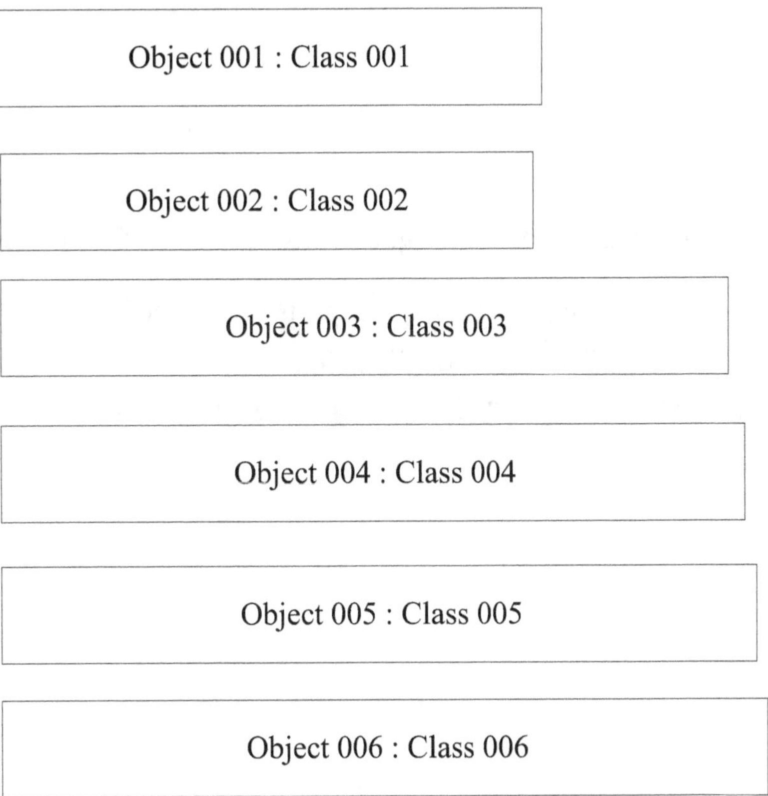

Figure 15-1. UML Object Diagram OD_{system}

15-2 OD Relation (ODR) of a System

In UML, the object diagram of a system OD_{system} is formally represented by a relation $ODR_{\text{system}} \subseteq \Gamma$, where Γ is a set of "objects", as shown in Figure 15-2.

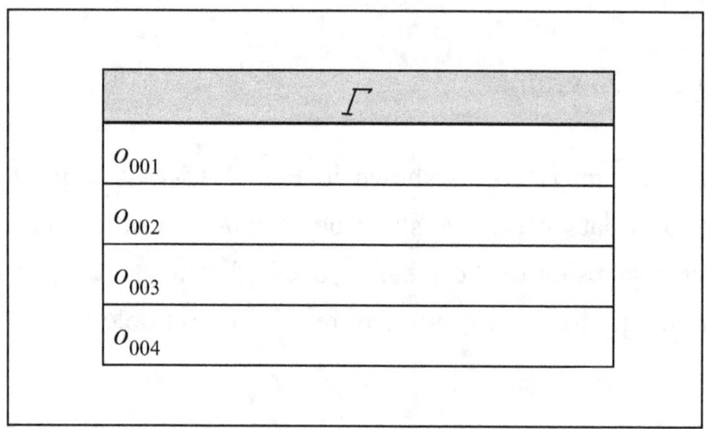

Figure 15-2. Relation ODR_{system}

15-3 Algorithm of Projecting an Object Diagram from O-M-SBC-PA

In O-M-SBC-PA, the process of a system is represented by a transition graph TG_{system} (defined as "$TG_1 \square TG_2 \; \square \ldots \square \; TG_m$") with the transition relation $TGR_{\text{system}} \subseteq \Psi_1 X \Delta X \Psi_2$ (defined as "$TGR_1 \square TGR_2 \square \ldots \square TGR_m$") as shown in Figure 15-3.

Ψ_1	Δ	Ψ_2
E_{11}	g_{11}	E_{12}
E_{12}	a_{12}	E_{13}
E_{13}	a_{13}	E_{14}
•	•	•
E_{1n}	a_{1n}	E_{11}

$\|$

Ψ_1	Δ	Ψ_2
E_{21}	g_{21}	E_{22}
E_{22}	a_{22}	E_{23}
E_{23}	a_{23}	E_{24}
•	•	•
E_{2n}	a_{2n}	E_{21}

$\|$ • • • • • • • • • $\|$

Ψ_1	Δ	Ψ_2
E_{m1}	g_{m1}	E_{m2}
E_{m2}	a_{m2}	E_{m3}
E_{m3}	a_{m3}	E_{m4}
•	•	•
E_{mn}	a_{mn}	E_{m1}

Figure 15-3. Relation TGR_{system}

We rewrite the TG relation of a system as $TGR_{\text{system}} \subseteq \Psi_1 \mathrm{X}\, N \,\mathrm{X}\, \Xi \,\mathrm{X}\, \Lambda \,\mathrm{X}\, \Theta \,\mathrm{X}\, \Gamma \,\mathrm{X}\, \Psi_2$ since the "type 1 or 2 interaction" is defined as a relation $\Lambda \subseteq N \,\mathrm{X}\, \Xi \,\mathrm{X}\, L \,\mathrm{X}\, \Gamma$ and the "operation call or operation return signature" is defined as a relation $L \subseteq \Lambda \mathrm{X}\, \Theta$.

Figure 15-4 shows the algorithm of projecting the OD relation $ODR_{system} \subseteq \Gamma$ from the TG relation $TGR_{system} \subseteq \Psi_1 X N X \Xi X \Lambda X \Theta X \Gamma X \Psi_2$.

```
For i = 1, m Loop
   SELECT Γ INTO ODR_i (Γ) FROM TGR_i ;
   INSERT INTO ODR_1~m (Γ) SELECT * FROM ODR_i;
End Loop;

SELECT DISTINCT * INTO ODR_system FROM ODR_1~m ;
```

Figure 15-4. Algorithm of Projecting the OD Relation from the TG Relation

Once we have the OD relation ODR_{system}, it is easy to get a UML object diagram of the system.

PART IV: CASE STUDY

Chapter 16: Online Shopping Systems

The online shopping system is a highly distributed world wide web-based system that provides services for purchasing items such as books or clothes. In the online shopping system, customers can request to order one or more items from the supplier. The customer provides personal details, such as address and credit card information. This information is stored in a customer account. If the credit card is valid, then a delivery order is created and sent to the supplier. The supplier checks the available inventory, confirms the order, and enters a planned shipping date. When the order is shipped, the customer is notified and the customer's credit card account is charged. The online shopping system also allows the customer to view the details of the delivery order.

Behaviors of the online shopping system consist of: a) *Make_Order_Request* behavior, b) *Confirm_Shipment_and_Bill_Customer* behavior and c) *View_Order* behavior.

In the *Make_Order_Request* behavior, the customer enters personal details. The system creates a customer account if one does not already exist. The customer's credit card is checked for validity and sufficient credit to pay for the requested catalog items. If the credit card check shows that the credit card is valid and has sufficient credit, then the customer purchase is approved and the system sends the purchase request to the supplier. In the *Confirm_Shipment_and_Bill_Customer* behavior, the supplier prepares the shipment manually and confirms that the shipment is ready for shipment. The system then retrieves the customer's credit card details from the customer account and bills the customer's credit card. In the *View_Order* behavior, the customer requests to view the details of the delivery order.

16-1 O-M-SBC-PA Process of the Online Shopping System

We first use O-M-SBC-PA to define the online shopping system. The O-M-SBC-PA process expression of the online shopping system, E_{OSS}, is defined as "$\mathbf{fix}(X_1=g_{11}\bullet v_{12}\bullet v_{13}\bullet v_{14}\bullet g_{15}\bullet g_{16}\bullet X_1)\square\mathbf{fix}(X_2=g_{21}\bullet v_{22}\bullet v_{23}\bullet v_{24}\bullet v_{25}\bullet v_{26}\bullet g_{27}\bullet X_2)\square\mathbf{fix}(X_3=g_{31}\bullet v_{32}\bullet v_{33}\bullet v_{34}\bullet g_{35}\bullet X_3)$".

16-2 O-M-SBC-PA Transition Graph of the Online Shopping System

We use the O-M-SBC-PA transition graph TG_{OSS} defined as "$TG_1\square TG_2\square TG_3$"

to represent the execution of the process expression of the online shopping system, as shown in Figure 16-1.

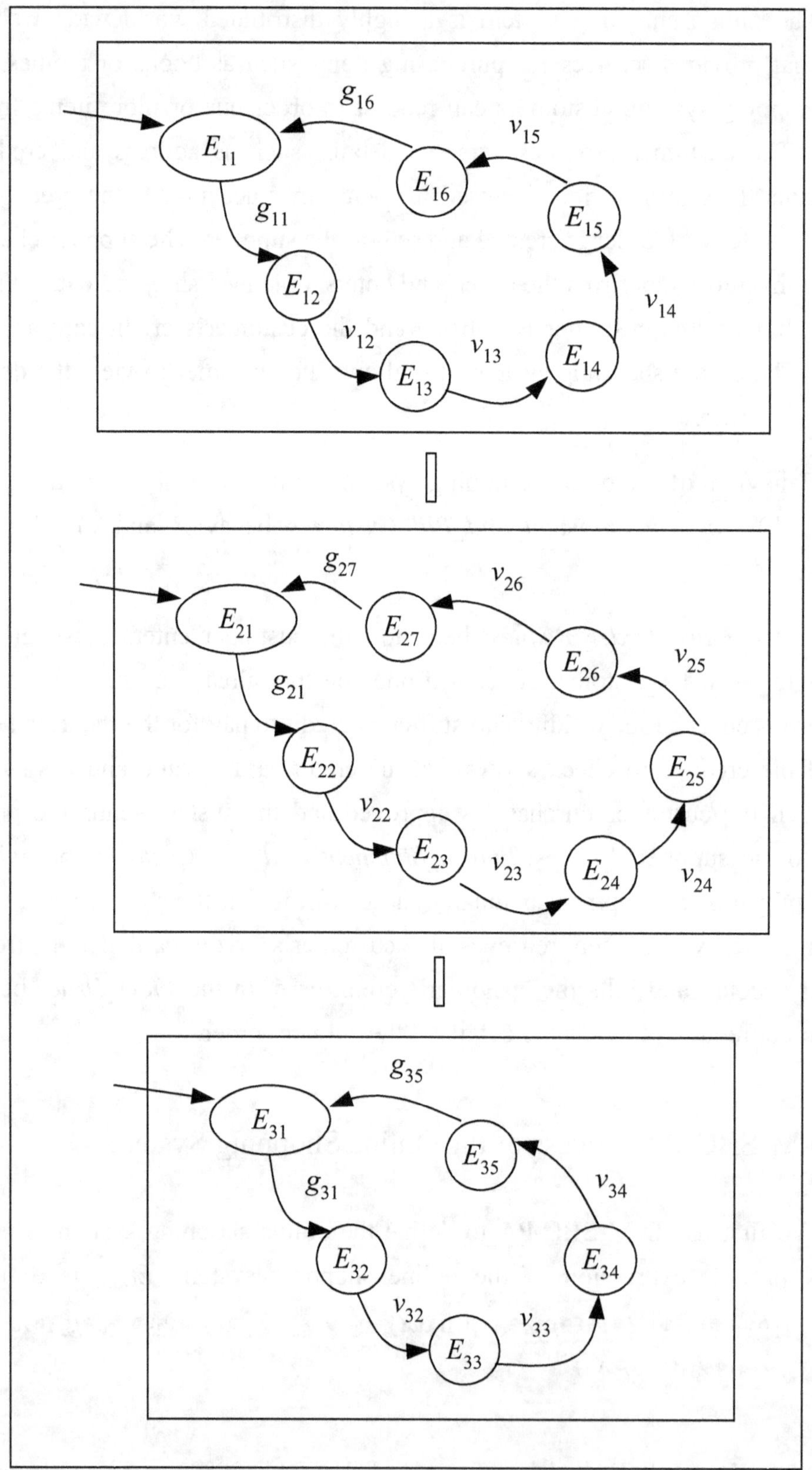

Figure 16-1. Transition Graph TG_{OSS}

16-3 TG Relation of the Online Shopping System

We use a TG relation $TGR_{OSS} \subseteq \Psi_1 \times N \times \Xi \times \Lambda \times \Theta \times \Gamma \times \Psi_2$ defined as "TGR_1□TGR_2□TGR_3" to represent the O-M-SBC-PA transition graph TG_{OSS} of the process expression of the online shopping system, as shown in Figure 16-2.

Ψ_1	Δ					Ψ_2
	N	Ξ	Λ	Θ	Γ	
	g_{11}					
E_{11}	*CAL*	Customer	Request_ Order_ from_ Customer	in Request_ Order_ Info	:Customer_ UI	E_{12}
	v_{12}					
E_{12}	*CAL*	:Customer_ UI	Request_ Order_ from_UI	in Request_ Order_ Info	:Customer_ Coordinator	E_{13}
	v_{13}					
E_{13}	*CAL*	:Customer_ Coordinator	Authorize_ Credit_ Card_ Charge	in Credit_ Card_Id; in Amount; out Authorization_ Response	:Credit_ Card_ Service	E_{14}
	v_{14}					
E_{14}	*CAL*	:Customer_ Coordinator	Store_ Order	in Order; out Order_Id	:Delivery_ Order_ Service	E_{15}
	v_{15}					
E_{15}	*RET*	:Customer_ UI	Request_ Order_ from_ UI	out Order_Info	:Customer_ Coordinator	E_{16}
	g_{16}					
E_{16}	*RET*	Customer	Request_ Order_ from_ Customer	out Order_Info	:Customer_ UI	E_{11}

Figure 16-2. Relation TGR_{OSS} (I)

Ψ_1	Δ					Ψ_2
	N	Ξ	Λ	Θ	Γ	
E_{21}			g_{21}			E_{22}
	CAL	Supplier	Shipping	in Order_Id	:Supplier_ UI	
E_{22}			v_{22}			E_{23}
	CAL	:Supplier_ UI	Ready_ for_ Shippment	in Order_Id	:Supplier_ Coordinator	
E_{23}			v_{23}			E_{24}
	CAL	:Supplier_ Coordinator	Request_ Invoice	in Order_Id; out Invoice	:Delivery_ Order_ Service	
E_{24}			v_{24}			E_{25}
	CAL	:Supplier_ Coordinator	Commit_ Credit_ Card_ Charge	in Credit_ Card_Id; in Amount; out Commit_ Response	:Credit_ Card_ Service	
E_{25}			v_{25}			E_{26}
	CAL	:Supplier_ Coordinator	Confirm_ Payment	in Order_Id; in Amount; out Order_ Status	:Delivery_ Order_ Service	
E_{26}			v_{26}			E_{27}
	RET	:Supplier_ UI	Ready_ for_ Shippment	out Order_ Status	:Supplier_ Coordinator	
E_{27}			g_{27}			E_{21}
	RET	Supplier	Shipping	out Order_ Status	:Supplier_ UI	

Figure 16-2. Relation TGR_{OSS} (II)

Ψ_1	Δ					Ψ_2
	N	Ξ	Λ	Θ	Γ	
E_{31}	g_{31}					E_{32}
	CAL	Customer	Request_ Order_ Status_ from_ Customer	in Order_Id	:Customer_ UI	
E_{32}	v_{32}					E_{33}
	CAL	:Customer_ UI	Request_ Order_ Status_ from_UI	in Order_Id	:Customer_ Coordinator	
E_{33}	v_{33}					E_{34}
	CAL	:Customer_ Coordinator	Read_Order	in Order_Id; out Order	:Delivery_ Order_ Service	
E_{34}	v_{34}					E_{35}
	RET	:Customer_ UI	Request_ Order_ Status_ from_UI	out Order_ Status	:Customer_ Coordinator	
E_{35}	g_{35}					E_{31}
	RET	Customer	Request_ Order_ Status_ from_ Customer	out Order_ Status	:Customer_ UI	

Figure 16-2. Relation TGR_{OSS} (III)

Chapter 17: Projecting a Use Case Diagram from the O-M-SBC-PA Transition Graph of the Online Shopping System

We use two steps to project a UML use case diagram from the O-M-SBC-PA transition graph of the online shopping system. First, we project the UCD relation from the transition graph of the Online Shopping System. Second, we will draw the UML use case diagram from the UCD relation of the online shopping system.

17-1 Projecting the UCD Relation from the Transition Graph of the Online Shopping System

The TG relation $TGR_{OSS} \subseteq \Psi_1 \text{ X } N \text{ X } \Xi \text{ X } \Lambda \text{ X } \Theta \text{ X } \Gamma \text{ X } \Psi_2$, defined as "$TGR_1$ ▯TGR_2▯TGR_3" and shown in Figure 16-2, is used to represent the O-M-SBC-PA transition graph TG_{OSS} of the online shopping system.

We apply the algorithm of projecting the UCD relation (i.e., $UCDR_{OSS}$) from the TG relation (i.e., TGR_{OSS}) of the online shopping system. After the projection, we get the relation $UCDR_{OSS} \subseteq B \text{X} U$ as shown in Figure 17-1.

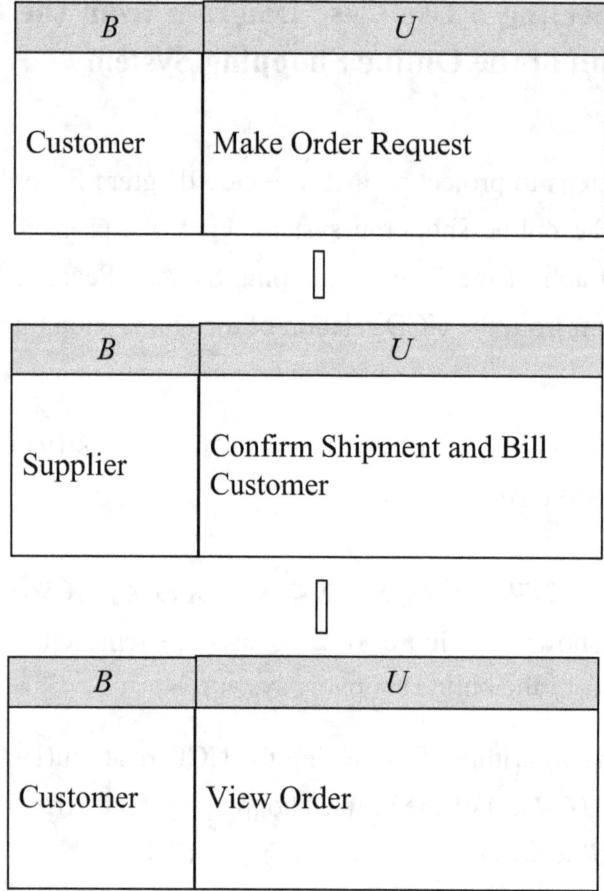

Figure 17-1. Relation $UCDR_{OSS}$

17-2 Achieving the Use Case Diagram from the UCD Relation of the Online Shopping System

From the UCD relation $UCDR_{OSS}$, we draw the corresponding UML use case diagram of the online shopping system, as shown in Figure 17-2.

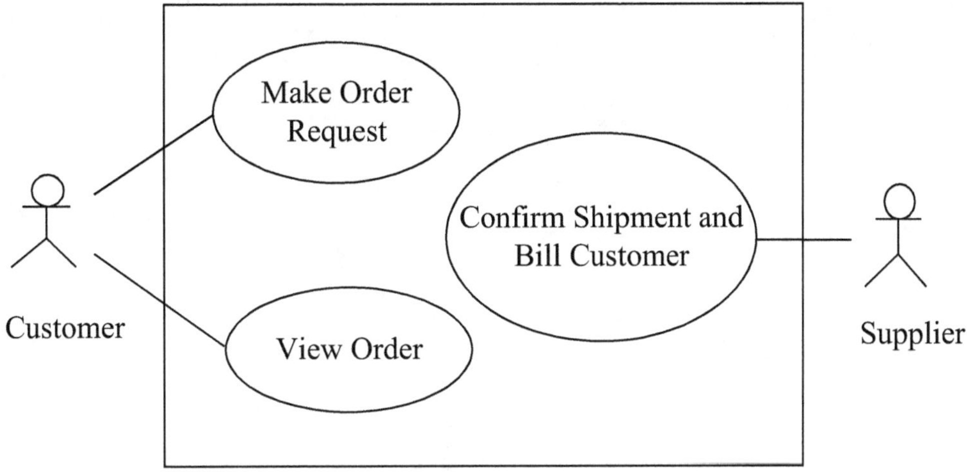

Figure 17-2. Use Case Diagram of the Online Shopping System

Chapter 18: Projecting a State Diagram from the O-M-SBC-PA Transition Graph of the Online Shopping System

We use two steps to project a UML state diagram from the O-M-SBC-PA transition graph of the online shopping system. First, we project the StD relation from the transition graph of the Online Shopping System. Second, we will draw the UML state diagram from the StD relation of the online shopping system.

18-1 Projecting the StD Relation from the Transition Graph of the Online Shopping System

The TG relation $TGR_{OSS} \subseteq \Psi_1 \text{ X } N \text{ X } \Xi \text{ X } \Lambda \text{ X } \Theta \text{ X } \Gamma \text{ X } \Psi_2$, defined as "$TGR_1$ $\square TGR_2 \square TGR_3$" and shown in Figure 16-2, is used to represent the O-M-SBC-PA transition graph TG_{OSS} of the online shopping system.

We apply the algorithm of projecting the StD relation (i.e., $StDR_{OSS}$) from the TG relation (i.e., TGR_{OSS}) of the online shopping system. After the projection, we get the relation $StDR_{OSS} \subseteq \Psi_1 X N X \Lambda X \Psi_2$ as shown in Figure 18-1.

Ψ_1	N	Λ	Ψ_2
E_{11}	*CAL*	Request_ Order_ from_ Customer	E_{12}
E_{12}	*CAL*	Request_ Order_ from_UI	E_{13}
E_{13}	*CAL*	Authorize_ Credit_ Card_ Charge	E_{14}
E_{14}	*CAL*	Store_ Order	E_{15}
E_{15}	*RET*	Request_ Order_ from_ UI	E_{16}
E_{16}	*RET*	Request_ Order_ from_ Customer	E_{11}

Figure 18-1. Relation $StDR_{\text{OSS}}$ (I)

Ψ_1	N	Λ	Ψ_1
E_{21}	*CAL*	Shipping	E_{22}
E_{22}	*CAL*	Ready_ for_ Shippment	E_{23}
E_{23}	*CAL*	Request_ Invoice	E_{24}
E_{24}	*CAL*	Commit_ Credit_ Card_ Charge	E_{25}
E_{25}	*CAL*	Confirm_ Payment	E_{26}
E_{26}	*RET*	Ready_ for_ Shippment	E_{27}
E_{27}	*RET*	Shipping	E_{21}

Figure 18-1. Relation $StDR_{OSS}$ (II)

Ψ_1	N	Λ	Ψ_1
E_{31}	*CAL*	Request_ Order_ Status_ from_ Customer	E_{32}
E_{32}	*CAL*	Request_ Order_ Status_ from_UI	E_{33}
E_{33}	*CAL*	Read_Order	E_{34}
E_{34}	*RET*	Request_ Order_ Status_ from_UI	E_{35}
E_{35}	*RET*	Request_ Order_ Status_ from_ Customer	E_{31}

Figure 18-1. Relation $StDR_{\mathrm{OSS}}$ (III)

18-2 Achieving the State Diagram from the StD Relation of the Online Shopping System

From the StD relation $StDR_{OSS}$, we draw the corresponding UML state diagram of the online shopping system, as shown in Figure 18-2.

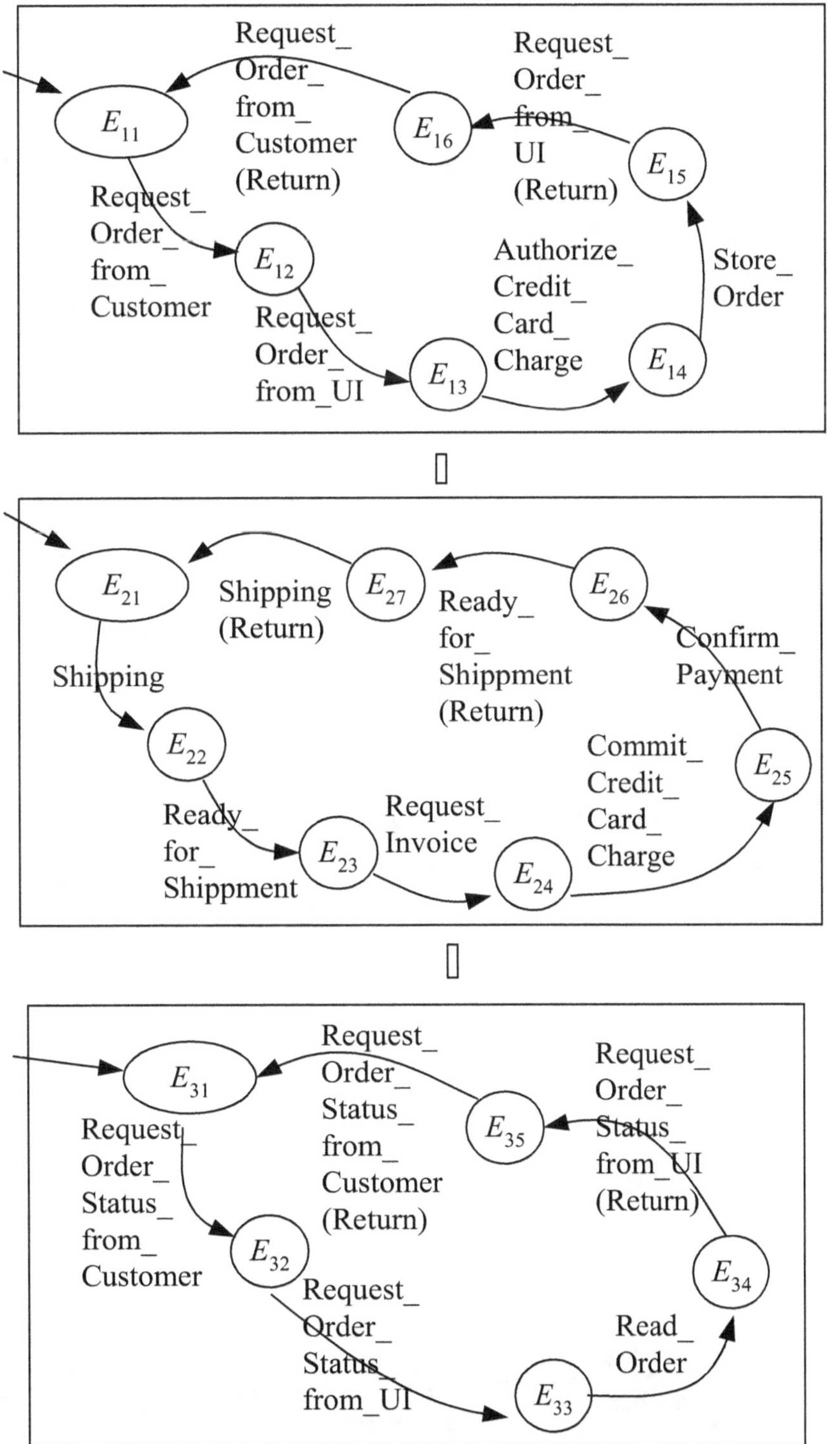

Figure 18-2. State Diagram of the Online Shopping System

[illegible].2 Achieving the State Diagram from the SD Relation of the Online Shopping System

From the SD relation SDR[illegible], we draw the corresponding UML state diagram of the online shopping system, as shown in Figure [illegible].

Figure [illegible] State Diagram of the Online Shopping System

Ψ_1	N	Λ	Ψ_1
E_{21}	*CAL*	Shipping	E_{22}
E_{22}	*CAL*	Ready_ for_ Shippment	E_{23}
E_{23}	*CAL*	Request_ Invoice	E_{24}
E_{24}	*CAL*	Commit_ Credit_ Card_ Charge	E_{25}
E_{25}	*CAL*	Confirm_ Payment	E_{26}
E_{26}	*RET*	Ready_ for_ Shippment	E_{27}
E_{27}	*RET*	Shipping	E_{21}

Figure 18-1. Relation $StDR_{\mathrm{OSS}}$ (II)

Ψ_1	N	Λ	Ψ_1
E_{31}	*CAL*	Request_Order_Status_from_Customer	E_{32}
E_{32}	*CAL*	Request_Order_Status_from_UI	E_{33}
E_{33}	*CAL*	Read_Order	E_{34}
E_{34}	*RET*	Request_Order_Status_from_UI	E_{35}
E_{35}	*RET*	Request_Order_Status_from_Customer	E_{31}

Figure 18-1. Relation $StDR_{\mathrm{OSS}}$ (III)

18-2 Achieving the State Diagram from the StD Relation of the Online Shopping System

From the StD relation $StDR_{OSS}$, we draw the corresponding UML state diagram of the online shopping system, as shown in Figure 18-2.

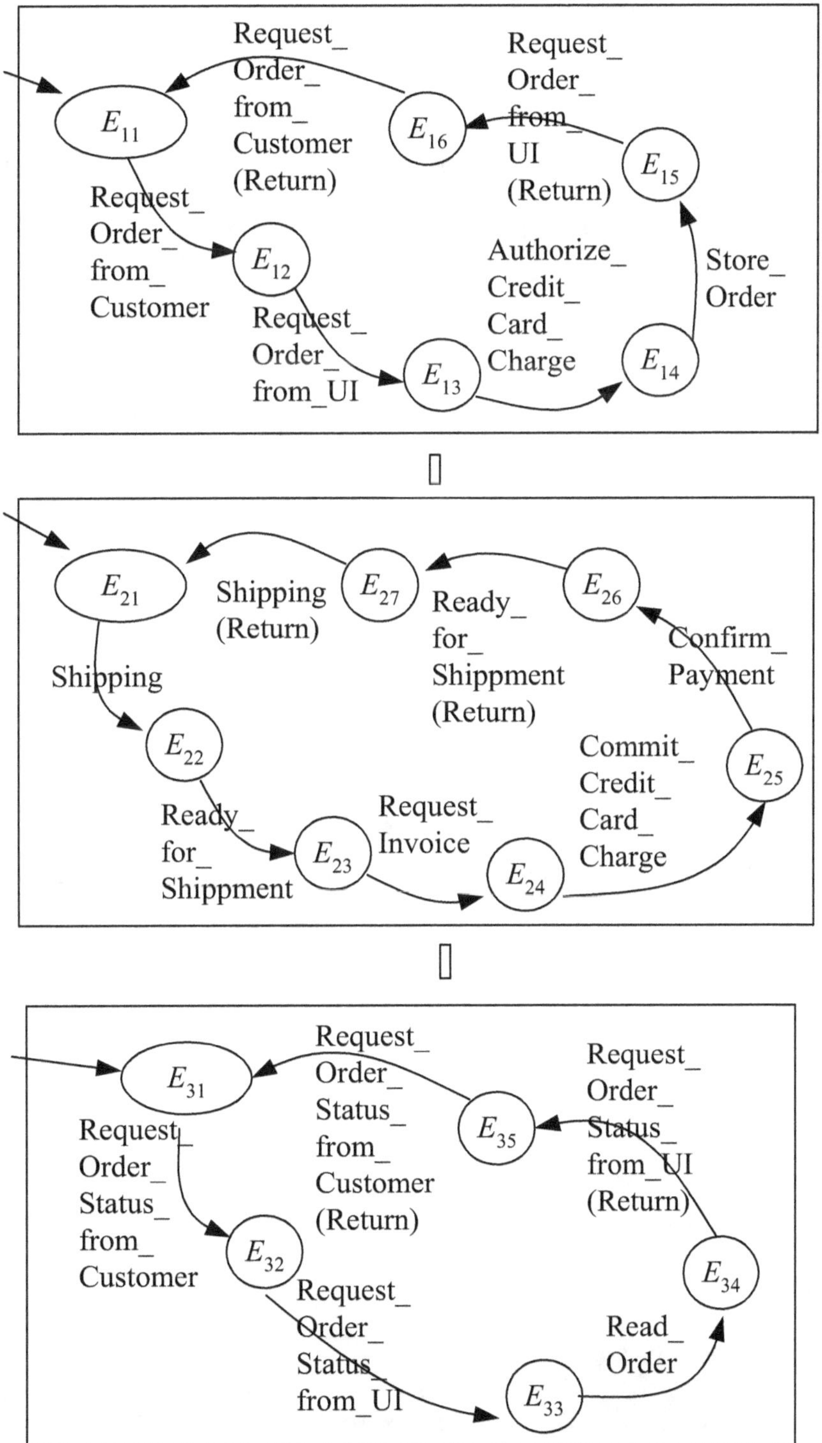

Figure 18-2. State Diagram of the Online Shopping System

Chapter 19: Projecting an Activity Diagram from the O-M-SBC-PA Transition Graph of the Online Shopping System

We use two steps to project a UML activity diagram from the O-M-SBC-PA transition graph of the online shopping system. First, we project the AD relation from the transition graph of the Online Shopping System. Second, we will draw the UML activity diagram from the AD relation of the online shopping system.

19-1 Projecting the AD Relation from the Transition Graph of the Online Shopping System

The TG relation $TGR_{OSS} \subseteq \Psi_1 \text{ X } N \text{ X } \Xi \text{ X } \Lambda \text{ X } \Theta \text{ X } \Gamma \text{ X } \Psi_2$, defined as "$TGR_1$ ▯TGR_2▯TGR_3" and shown in Figure 16-2, is used to represent the O-M-SBC-PA transition graph TG_{OSS} of the online shopping system.

We apply the algorithm of projecting the AD relation (i.e., ADR_{OSS}) from the TG relation (i.e., TGR_{OSS}) of the online shopping system. After the projection, we get the relation $ADR_{OSS} \subseteq \Psi_1 \text{X} N \text{X} \Lambda \text{X} \Psi_2$ as shown in Figure 19-1.

Ψ_1	*N*	Λ	Ψ_2
E_{11}	*CAL*	Request_ Order_ from_ Customer	E_{12}
E_{12}	*CAL*	Request_ Order_ from_UI	E_{13}
E_{13}	*CAL*	Authorize_ Credit_ Card_ Charge	E_{14}
E_{14}	*CAL*	Store_ Order	E_{15}
E_{15}	*RET*	Request_ Order_ from_ UI	E_{16}
E_{16}	*RET*	Request_ Order_ from_ Customer	E_{11}

Figure 19-1. Relation ADR_{OSS} (I)

Ψ_1	N	Λ	Ψ_1
E_{21}	CAL	Shipping	E_{22}
E_{22}	CAL	Ready_ for_ Shippment	E_{23}
E_{23}	CAL	Request_ Invoice	E_{24}
E_{24}	CAL	Commit_ Credit_ Card_ Charge	E_{25}
E_{25}	CAL	Confirm_ Payment	E_{26}
E_{26}	RET	Ready_ for_ Shippment	E_{27}
E_{27}	RET	Shipping	E_{21}

Figure 19-1. Relation ADR_{OSS} (II)

Ψ_1	N	Λ	Ψ_1
E_{31}	*CAL*	Request_ Order_ Status_ from_ Customer	E_{32}
E_{32}	*CAL*	Request_ Order_ Status_ from_UI	E_{33}
E_{33}	*CAL*	Read_Order	E_{34}
E_{34}	*RET*	Request_ Order_ Status_ from_UI	E_{35}
E_{35}	*RET*	Request_ Order_ Status_ from_ Customer	E_{31}

Figure 19-1. Relation ADR_{OSS} (III)

19-2 Achieving the Activity Diagram from the AD Relation of the Online Shopping System

From the AD relation ADR_{OSS}, we draw the corresponding UML activity diagram of the online shopping system, as shown in Figure 19-2.

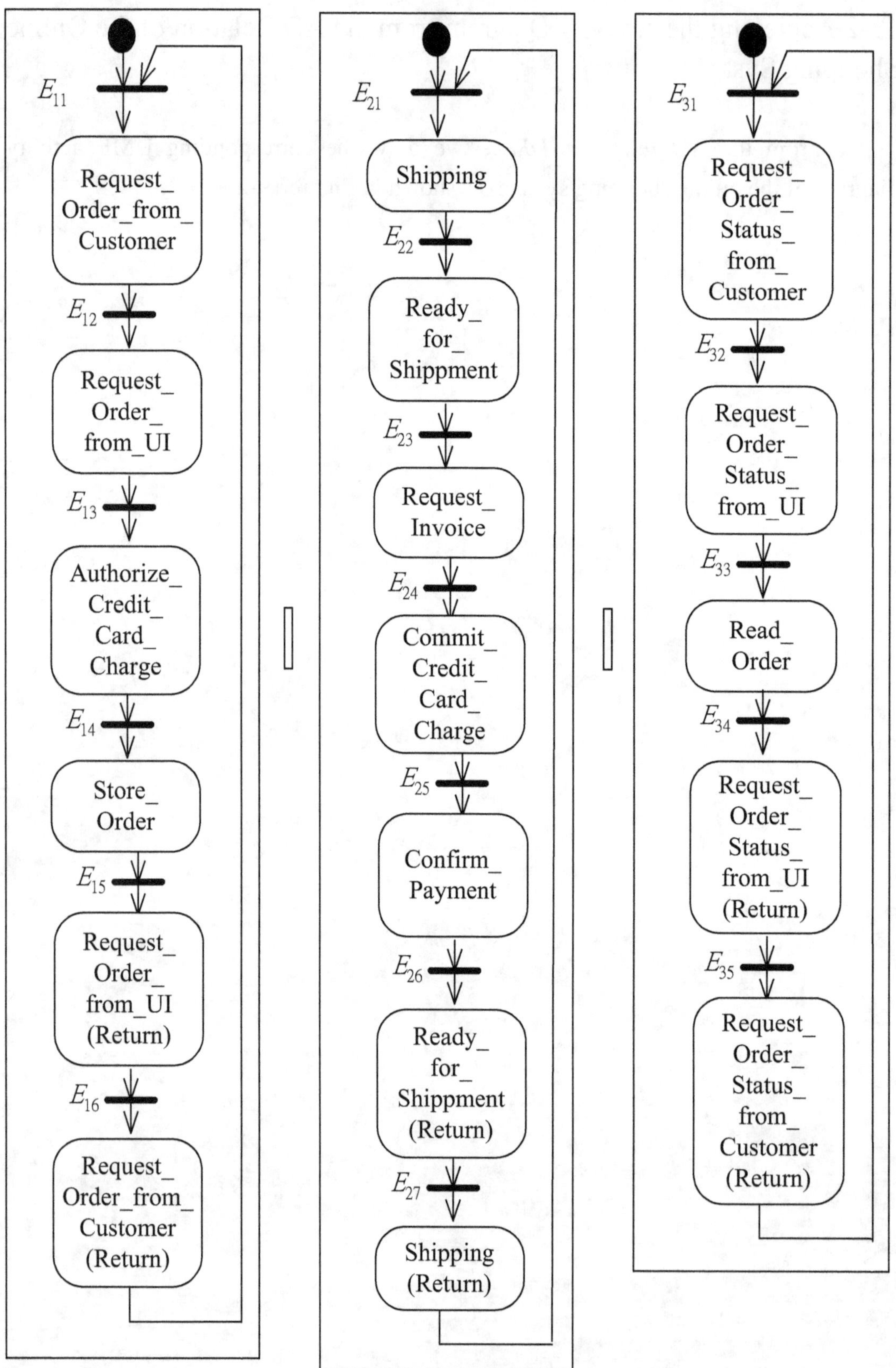

Figure 19-2. UML Activity Diagram of the Online Shopping System

Chapter 20: Projecting a Sequence Diagram from the O-M-SBC-PA Transition Graph of the Online Shopping System

We use two steps to project a UML sequence diagram from the O-M-SBC-PA transition graph of the online shopping system. First, we project the SqD relation from the transition graph of the Online Shopping System. Second, we will draw the UML sequence diagram from the SqD relation of the online shopping system.

20-1 Projecting the SqD Relation from the Transition Graph of the Online Shopping System

The TG relation $TGR_{OSS} \subseteq \Psi_1 \text{ X } N \text{ X } \Xi \text{ X } \Lambda \text{ X } \Theta \text{ X } \Gamma \text{ X } \Psi_2$, defined as "$TGR_1$ ▯ TGR_2 ▯ TGR_3" and shown in Figure 16-2, is used to represent the O-M-SBC-PA transition graph TG_{OSS} of the online shopping system.

We apply the algorithm of projecting the SqD relation (i.e., $SqDR_{OSS}$) from the TG relation (i.e., TGR_{OSS}) of the online shopping system. After the projection, we get the relation $SqDR_{OSS} \subseteq E \text{ X } N \text{ X } \Xi \text{ X } \Lambda \text{ X } \Theta \text{ X } \Gamma$ as shown in Figure 20-1.

E	N	Ξ	Λ	Θ	Γ
1	*CAL*	Customer	Request_ Order_ from_ Customer	in Request_ Order_ Info	:Customer_ UI
2	*CAL*	:Customer_ UI	Request_ Order_ from_UI	in Request_ Order_ Info	:Customer_ Coordinator
3	*CAL*	:Customer_ Coordinator	Authorize_ Credit_ Card_ Charge	in Credit_ Card_Id; in Amount; out Authorization_ Response	:Credit_ Card_ Service
4	*CAL*	:Customer_ Coordinator	Store_ Order	in Order; out Order_Id	:Delivery_ Order_ Service
5	*RET*	:Customer_ UI	Request_ Order_ from_ UI	out Order_Info	:Customer_ Coordinator
6	*RET*	Customer	Request_ Order_ from_ Customer	out Order_Info	:Customer_ UI

Figure 20-1. Relation $SqDR_{\mathrm{OSS}}$ (I)

E	N	Ξ	Λ	Θ	Γ
1	*CAL*	Supplier	Shipping	in Order_Id	:Supplier_ UI
2	*CAL*	:Supplier_ UI	Ready_ for_ Shippment	in Order_Id	:Supplier_ Coordinator
3	*CAL*	:Supplier_ Coordinator	Request_ Invoice	in Order_Id; out Invoice	:Delivery_ Order_ Service
4	*CAL*	:Supplier_ Coordinator	Commit_ Credit_ Card_ Charge	in Credit_ Card_Id; in Amount; out Commit_ Response	:Credit_ Card_ Service
5	*CAL*	:Supplier_ Coordinator	Confirm_ Payment	in Order_Id; in Amount; out Order_ Status	:Delivery_ Order_ Service
6	*RET*	:Supplier_ UI	Ready_ for_ Shippment	out Order_ Status	:Supplier_ Coordinator
7	*RET*	Supplier	Shipping	out Order_ Status	:Supplier_ UI

Figure 20-1. Relation $SqDR_{OSS}$ (II)

E	N	Ξ	Λ	Θ	Γ
1	*CAL*	Customer	Request_ Order_ Status_ from_ Customer	in Order_Id	:Customer_ UI
2	*CAL*	:Customer_ UI	Request_ Order_ Status_ from_UI	in Order_Id	:Customer_ Coordinator
3	*CAL*	:Customer_ Coordinator	Read_Order	in Order_Id; out Order	:Delivery_ Order_ Service
4	*RET*	:Customer_ UI	Request_ Order_ Status_ from_UI	out Order_ Status	:Customer_ Coordinator
5	*RET*	Customer	Request_ Order_ Status_ from_ Customer	out Order_ Status	:Customer_ UI

Figure 20-1. Relation $SqDR_{\mathrm{OSS}}$ (III)

20-2 Achieving the Sequence Diagram from the SqD Relation of the Online Shopping System

From the SqD relation $SqDR_{OSS}$, we draw the corresponding UML sequence diagram of the online shopping system, as shown in Figure 20-2.

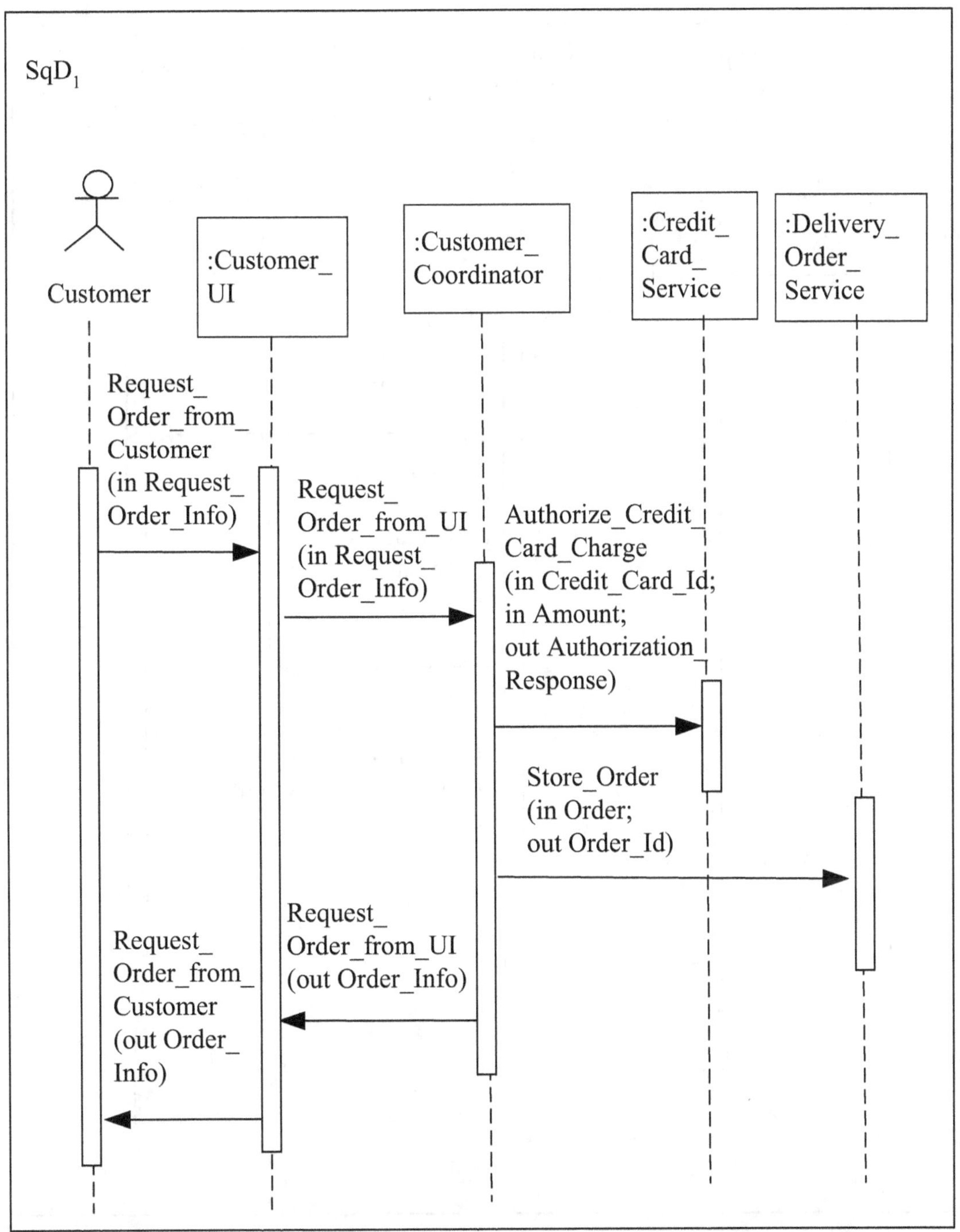

Figure 20-2. Sequence Diagram of the Online Shopping System (I)

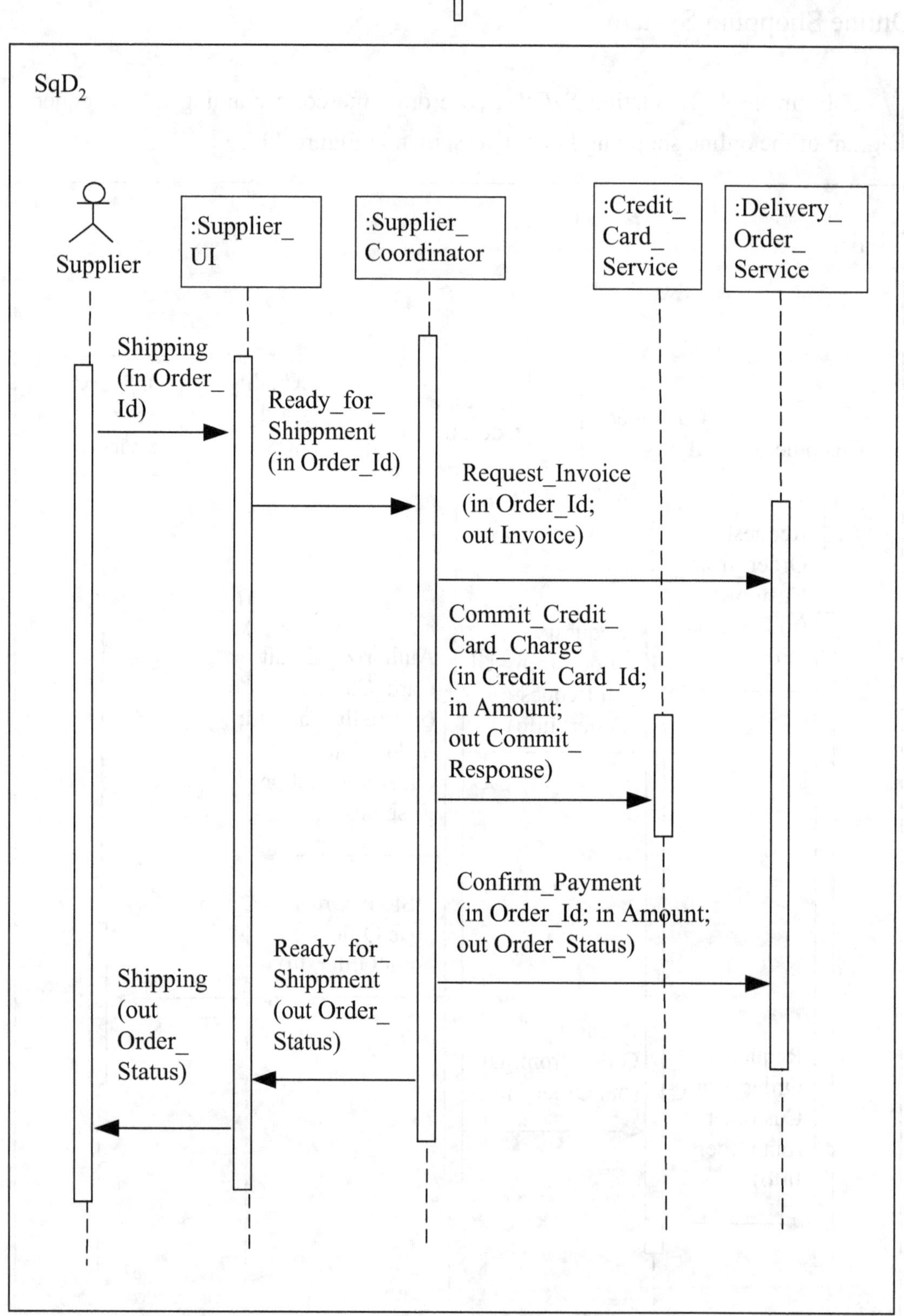

Figure 20-2. Sequence Diagram of the Online Shopping System (II)

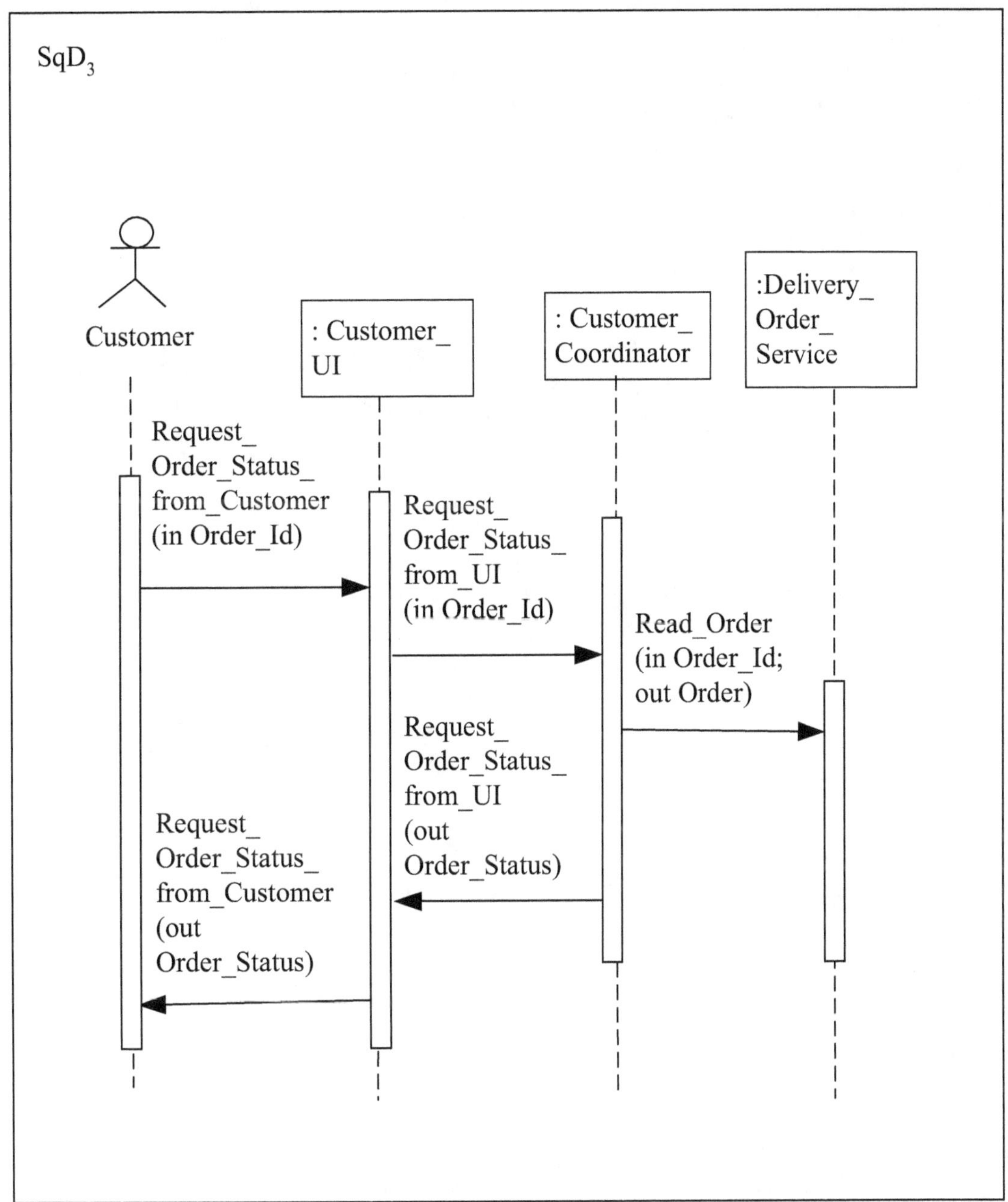

Figure 20-2. Sequence Diagram of the Online Shopping System (III)

Chapter 21: Projecting a Communication Diagram from the O-M-SBC-PA Transition Graph of the Online Shopping System

We use two steps to project a UML communication diagram from the O-M-SBC-PA transition graph of the online shopping system. First, we project the ComD relation from the transition graph of the Online Shopping System. Second, we will draw the UML communication diagram from the ComD relation of the online shopping system.

21-1 Projecting the ComD Relation from the Transition Graph of the Online Shopping System

The TG relation $TGR_{OSS} \subseteq \Psi_1 \text{ X } N \text{ X } \Xi \text{ X } \Lambda \text{ X } \Theta \text{ X } \Gamma \text{ X } \Psi_2$, defined as "$TGR_1$ $\square TGR_2 \square TGR_3$" and shown in Figure 16-2, is used to represent the O-M-SBC-PA transition graph TG_{OSS} of the online shopping system.

We apply the algorithm of projecting the ComD relation (i.e., $ComDR_{OSS}$) from the TG relation (i.e., TGR_{OSS}) of the online shopping system. After the projection, we get the relation $ComDR_{OSS} \subseteq E \text{ X } N \text{ X } \Xi \text{ X } \Lambda \text{ X } \Theta \text{ X } \Gamma$ as shown in Figure 21-1.

E	N	Ξ	Λ	Θ	Γ
1	*CAL*	Customer	Request_ Order_ from_ Customer	in Request_ Order_ Info	:Customer_ UI
2	*CAL*	:Customer_ UI	Request_ Order_ from_UI	in Request_ Order_ Info	:Customer_ Coordinator
3	*CAL*	:Customer_ Coordinator	Authorize_ Credit_ Card_ Charge	in Credit_ Card_Id; in Amount; out Authorization_ Response	:Credit_ Card_ Service
4	*CAL*	:Customer_ Coordinator	Store_ Order	in Order; out Order_Id	:Delivery_ Order_ Service
5	*RET*	:Customer_ UI	Request_ Order_ from_ UI	out Order_Info	:Customer_ Coordinator
6	*RET*	Customer	Request_ Order_ from_ Customer	out Order_Info	:Customer_ UI

Figure 21-1. Relation $ComDR_{\mathrm{OSS}}$ (I)

▯

E	*N*	*Ξ*	*Λ*	*Θ*	*Γ*
1	*CAL*	Supplier	Shipping	in Order_Id	:Supplier_ UI
2	*CAL*	:Supplier_ UI	Ready_ for_ Shippment	in Order_Id	:Supplier_ Coordinator
3	*CAL*	:Supplier_ Coordinator	Request_ Invoice	in Order_Id; out Invoice	:Delivery_ Order_ Service
4	*CAL*	:Supplier_ Coordinator	Commit_ Credit_ Card_ Charge	in Credit_ Card_Id; in Amount; out Commit_ Response	:Credit_ Card_ Service
5	*CAL*	:Supplier_ Coordinator	Confirm_ Payment	in Order_Id; in Amount; out Order_ Status	:Delivery_ Order_ Service
6	*RET*	:Supplier_ UI	Ready_ for_ Shippment	out Order_ Status	:Supplier_ Coordinator
7	*RET*	Supplier	Shipping	out Order_ Status	:Supplier_ UI

Figure 21-1. Relation $ComDR_{\mathrm{OSS}}$ (II)

▯

E	N	Ξ	Λ	Θ	Γ
1	*CAL*	Customer	Request_Order_Status_from_Customer	in Order_Id	:Customer_UI
2	*CAL*	:Customer_UI	Request_Order_Status_from_UI	in Order_Id	:Customer_Coordinator
3	*CAL*	:Customer_Coordinator	Read_Order	in Order_Id; out Order	:Delivery_Order_Service
4	*RET*	:Customer_UI	Request_Order_Status_from_UI	out Order_Status	:Customer_Coordinator
5	*RET*	Customer	Request_Order_Status_from_Customer	out Order_Status	:Customer_UI

Figure 21-1. Relation $ComDR_{OSS}$ (III)

21-2 Achieving the Communication Diagram from the ComD Relation of the Online Shopping System

From the ComD relation *ComDR*$_{OSS}$, we draw the corresponding UML communication diagram of the online shopping system, as shown in Figure 21-2.

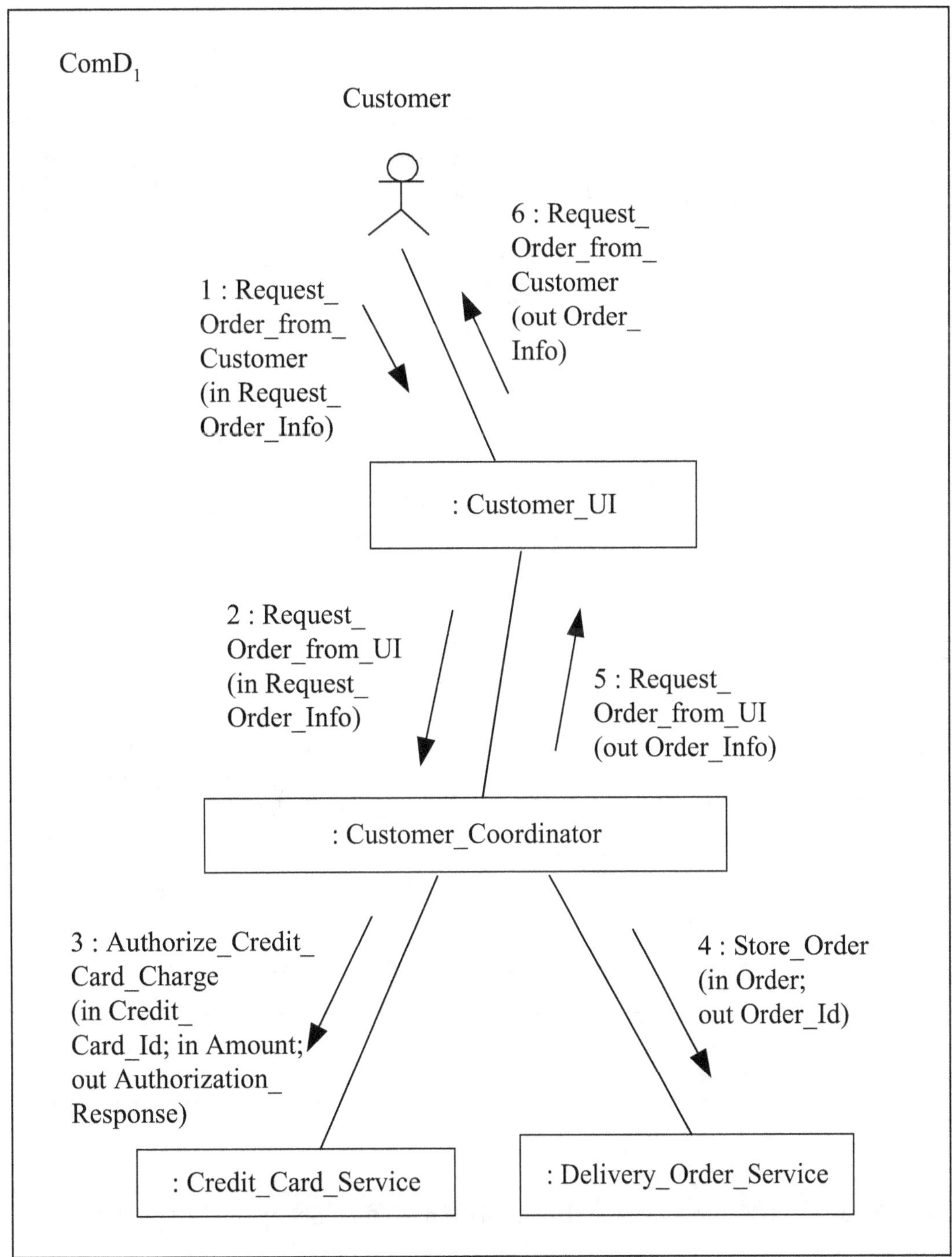

Figure 21-2. Communication Diagram of the Online Shopping System (I)

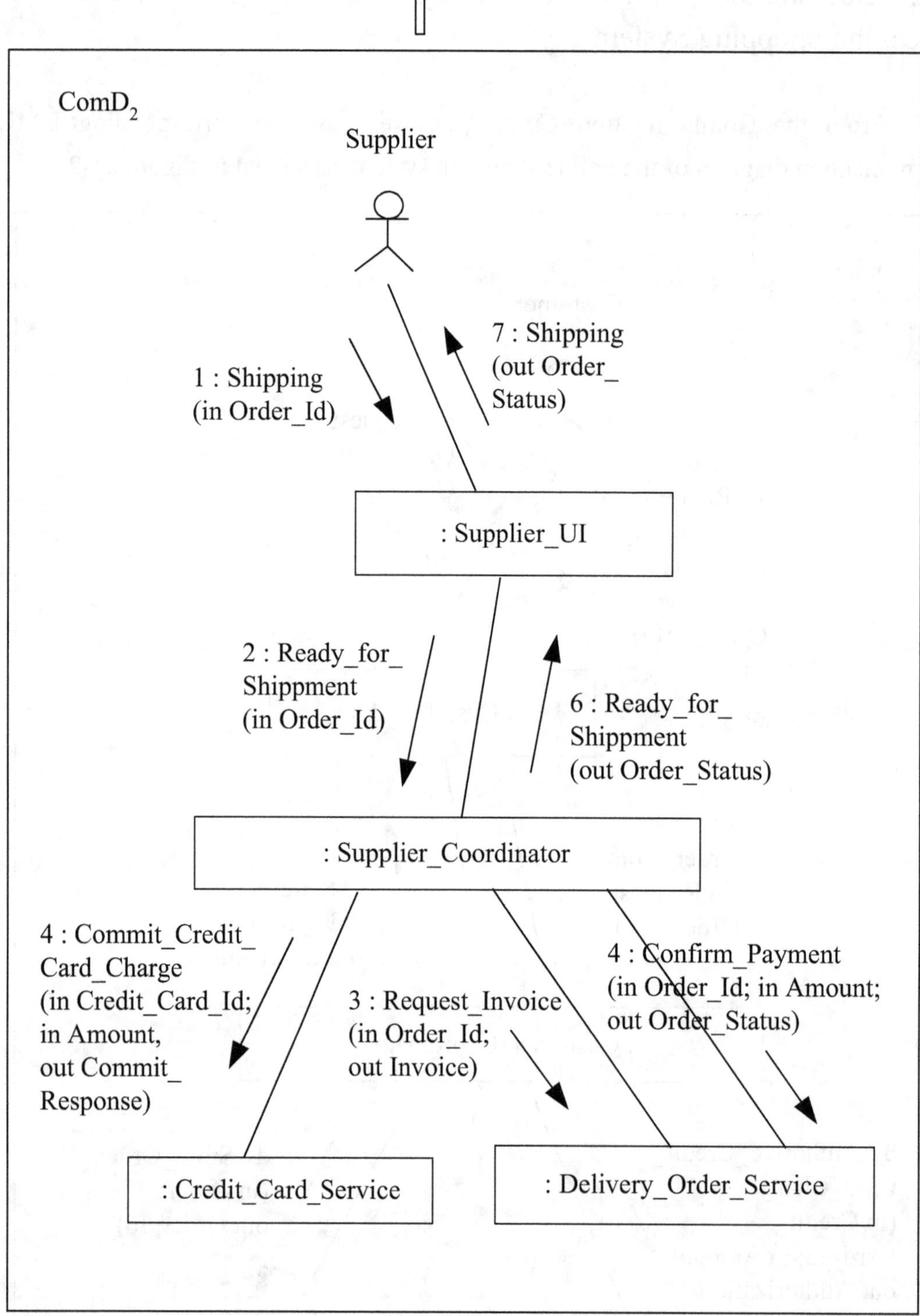

Figure 21-2. Communication Diagram of the Online Shopping System (II)

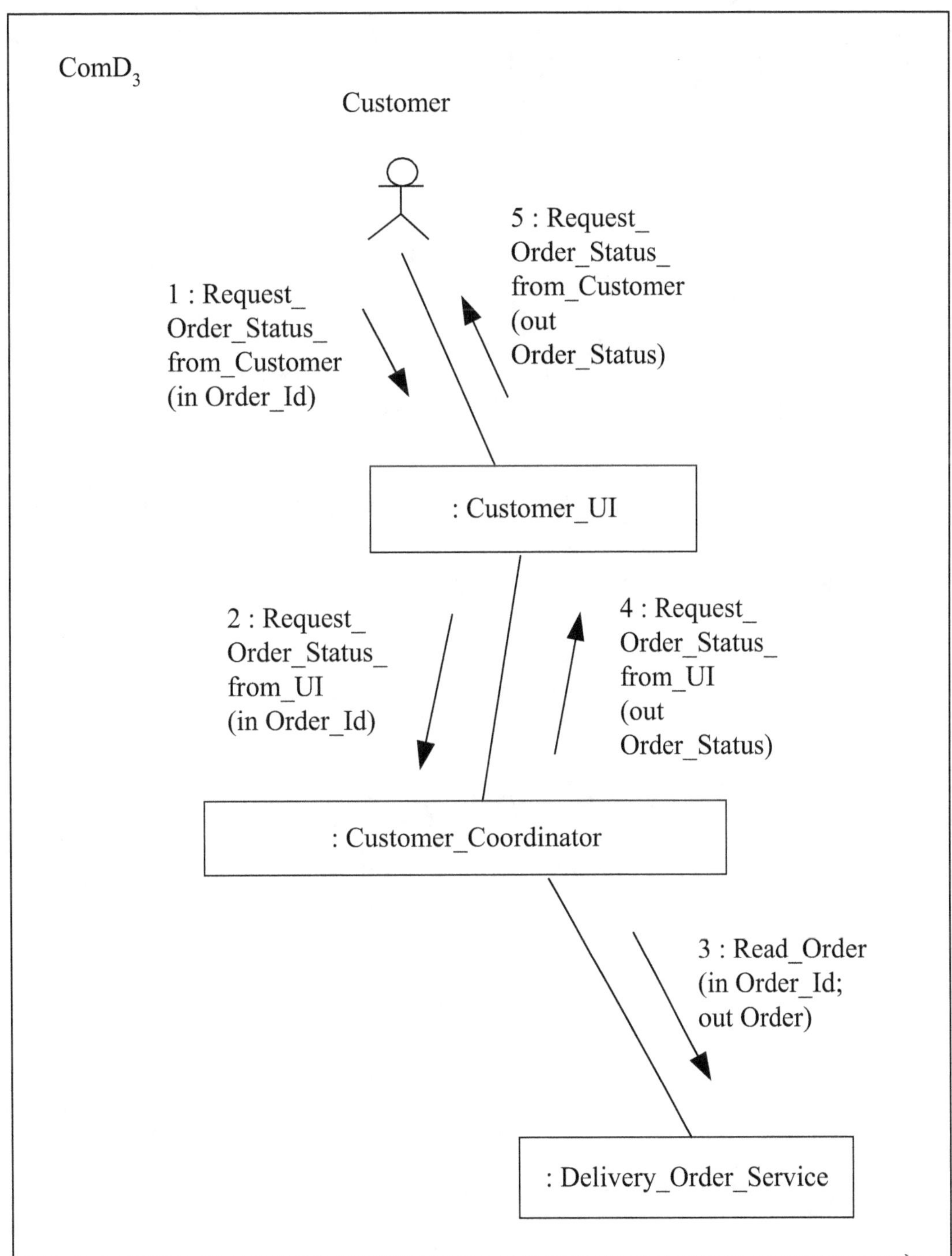

Figure 21-2. Communication Diagram of the Online Shopping System (III)

Chapter 22: Projecting a Class Diagram from the O-M-SBC-PA Transition Graph of the Online Shopping System

We use two steps to project a UML class diagram from the O-M-SBC-PA transition graph of the online shopping system. First, we project the ClsD relation from the transition graph of the Online Shopping System. Second, we will draw the UML class diagram from the ClsD relation of the online shopping system.

22-1 Projecting the ClsD Relation from the Transition Graph of the Online Shopping System

The TG relation $TGR_{OSS} \subseteq \Psi_1 \mathrm{X}\, N \mathrm{X}\, \Xi\, \mathrm{X}\, \Lambda\, \mathrm{X}\, \Theta\, \mathrm{X}\, \Gamma\, \mathrm{X}\, \Psi_2$, defined as "$TGR_1$ TGR_2 TGR_3" and shown in Figure 16-2, is used to represent the O-M-SBC-PA transition graph TG_{OSS} of the online shopping system.

We apply the algorithm of projecting the ClsD relation (i.e., $ClsDR_{OSS}$) from the TG relation (i.e., TGR_{OSS}) of the online shopping system. After the projection, we get the relation $ClsDR_{OSS} \subseteq C \mathrm{X}\, \Lambda\, \mathrm{X}\, \Theta$ as shown in Figure 22-1.

C	*Λ*	*Θ*
Customer_UI	Request_Order_ from_Customer	in Request_Order_Info; out Order_Info
Customer_UI	Request_Order_Status_ from_Customer	in Order_Id; out Order_Status
Supplier_UI	Shipping	in Order_Id; out Order_Status
Customer_Coordinator	Request_Order_ from_UI	in Request_Order_Info; out Order_Info
Customer_Coordinator	Request_Order_ Status_from_UI	in Order_Id; out Order_Status
Supplier_Coordinator	Ready_for_Shippment	in Order_Id; out Order_Status
Credit_Card_Service	Authorize_Credit_ Card_Charge	in Credit_Card_Id; in Amount; out Authorization_Response
Credit_Card_Service	Commit_Credit_ Card_Charge	in Credit_Card_Id; in Amount; out Commit_Response
Delivery_Order_Service	Store_Order	in Order; out Order_Id
Delivery_Order_Service	Request_Invoice	in Order_Id; out Invoice
Delivery_Order_Service	Confirm_Payment	in Order_Id; in Amount; out Order_Status
Delivery_Order_Service	Read_Order	in Order_Id; out Order

Figure 22-1. Relation $ClsDR_{\mathrm{OSS}}$

22-2 Achieving the Class Diagram from the ClsD Relation of the Online Shopping System

From the ClsD relation $ClsDR_{OSS}$, we draw the corresponding UML class diagram of the online shopping system, as shown in Figure 22-2.

Customer_UI
Request_Order_from_Customer (in Request_Order_Info; out Order_Info)
Request_Order_Status_from_Customer (in Order_Id; out Order_Status)

Supplier_UI
Shipping (in Order_Id; out Order_Status)

Supplier_Coordinator
Ready_for_Shippment (in Order_Id; out Order_Status)

Customer_Coordinator
Request_Order_from_UI (in Request_Order_Info; out Order_Info)
Request_Order_Status_from_UI (in Order_Id; out Order_Status)

Delivery_Order_Service
Store_Order (in Order; out Order_Id)
Request_Invoice (in Order_Id; out Invoice)
Confirm_Payment (in Order_Id; in Amount; out Order_Status)
Read_Order (in Order_Id; out Order)

Credit_Card_Service
Authorize_Credit_Card_Charge (in Credit_Card_Id; in Amount; out Authorization_Response)
Commit_Credit_Card_Charge (in Credit_Card_Id; in Amount; out Commit_Response)

Figure 22-2. Class Diagram of the Online Shopping System

[illegible]

Chapter 23: Projecting an Object Diagram from the O-M-SBC-PA Transition Graph of the Online Shopping System

We use two steps to project a UML object diagram from the O-M-SBC-PA transition graph of the online shopping system. First, we project the OD relation from the transition graph of the Online Shopping System. Second, we will draw the UML object diagram from the OD relation of the online shopping system.

23-1 Projecting the OD Relation from the Transition Graph of the Online Shopping System

The TG relation $TGR_{\mathrm{OSS}} \subseteq \Psi_1 \times N \times \Xi \times \Lambda \times \Theta \times \Gamma \times \Psi_2$, defined as "$TGR_1$ ▯TGR_2▯TGR_3" and shown in Figure 16-2, is used to represent the O-M-SBC-PA transition graph TG_{OSS} of the online shopping system.

We apply the algorithm of projecting the OD relation (i.e., ODR_{OSS}) from the TG relation (i.e., TGR_{OSS}) of the online shopping system. After the projection, we get the relation $ODR_{\mathrm{OSS}} \subseteq \Gamma$ as shown in Figure 23-1.

Γ
: Customer_UI
: Supplier_UI
: Customer_Coordinator
: Supplier_Coordinator
: Credit_Card_Service
: Delivery_Order_Service

Figure 23-1. Relation ODR_{OSS}

23-2 Achieving the Object Diagram from the OD Relation of the Online Shopping System

From the OD relation ODR_{OSS}, we draw the corresponding UML object diagram of the online shopping system, as shown in Figure 23-2.

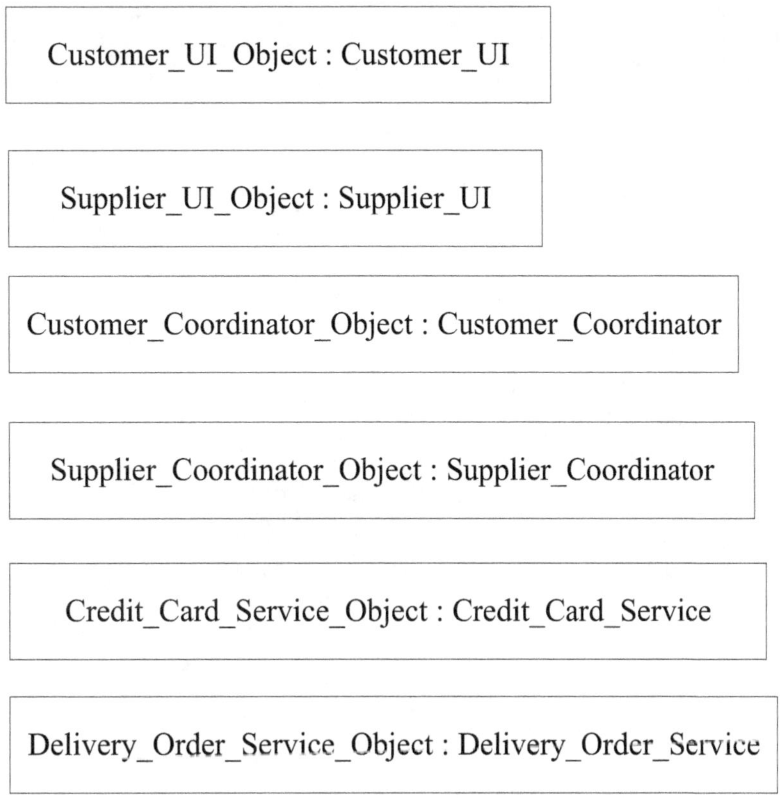

Customer_UI_Object : Customer_UI

Supplier_UI_Object : Supplier_UI

Customer_Coordinator_Object : Customer_Coordinator

Supplier_Coordinator_Object : Supplier_Coordinator

Credit_Card_Service_Object : Credit_Card_Service

Delivery_Order_Service_Object : Delivery_Order_Service

Figure 23-2. Object Diagram of the Online Shopping System

BIBLIOGRAPHY

[Alla15] Allaki, D. et al., "A New Taxonomy of Inconsistencies in UML Models with their Detection Methods for better MDE", *International Journal of Computer Science and Applications*, 12(1), pp. 48-65, 2015.

[Bash16] R. S. Bashir, R. S. et al., "UML Models Consistency Management: Guidelines for Software Quality Manager", *International Journal of Information Management*, 2016.

[Berg87] Bergstra, J. A. et al., "ACPτ: A Universal Axiom System for Process Specification," *CWI Quarterly* 15, 1987, pp. 3-23.

[Blah04] Blaha, M. R. et al., *Object-Oriented Modeling and Design with UML*, 2nd Edition, Pearson, 2004.

[Bram17] Brambilla, M. et al., Model-Driven Software Engineering in Practice, 2nd Edition, Pearson, Morgan & Claypool, 2017.

[Burd10] Burd, S. D., *Systems Architecture*, 6th Edition, Cengage Learning, 2010.

[Chao14a] Chao, W. S., *Systems Thingking 2.0: Architectural Thinking Using the SBC Architecture Description Language*, CreateSpace Independent Publishing Platform, 2014.

[Chao14b] Chao, W. S., *General Systems Theory 2.0: General Architectural Theory Using the SBC Architecture*, CreateSpace Independent Publishing Platform, 2014.

[Chao14c] Chao, W. S., *Software Modeling and Architecting: Structure-Behavior Coalescence for Software Architecture*, CreateSpace Independent Publishing Platform, 2014.

[Chao15a] Chao, W. S., *A Process Algebra For Systems Architecture: The Structure-*

Behavior Coalescence Approach, CreateSpace Independent Publishing Platform, 2015.

[Chao15b] Chao, W. S., *An Observation Congruence Model For Systems Architecture: The Structure-Behavior Coalescence Approach*, CreateSpace Independent Publishing Platform, 2015.

[Chao16] Chao, W. S., *System: Contemporary Concept, Definition, and Language*, CreateSpace Independent Publishing Platform, 2016.

[Chao17a] Chao, W. S., *Channel-Based Single-Queue SBC Process Algebra For Systems Definition: General Architectural Theory at Work*, CreateSpace Independent Publishing Platform, 2017.

[Chao17b] Chao, W. S., *Channel-Based Multi-Queue SBC Process Algebra For Systems Definition: General Architectural Theory at Work*, CreateSpace Independent Publishing Platform, 2017.

[Chao17c] Chao, W. S., *Channel-Based Infinite-Queue SBC Process Algebra For Systems Definition: General Architectural Theory at Work*, CreateSpace Independent Publishing Platform, 2017.

[Chao17d] Chao, W. S., *Operation-Based Single-Queue SBC Process Algebra For Systems Definition: General Architectural Theory at Work*, CreateSpace Independent Publishing Platform, 2017.

[Chao17e] Chao, W. S., *Operation-Based Multi-Queue SBC Process Algebra For Systems Definition: Unification of Systems Structure and Systems Behavior*, CreateSpace Independent Publishing Platform, 2017.

[Chao17f] Chao, W. S., *Operation-Based Infinite-Queue SBC Process Algebra For Systems Definition: Unification of Systems Structure and Systems Behavior*, CreateSpace Independent Publishing Platform, 2017.

[Chec99] Checkland, P., *Systems Thinking, Systems Practice: Includes a 30-Year*

Retrospective, 1st Edition, Wiley, 1999.

[Craw15] Crawley, P. et al., *System Architecture: Strategy and Product Development for Complex Systems*, Prentice Hall, 2015.

[Dam06] Dam, S., *DoD Architecture Framework: A Guide to Applying System Engineering to Develop Integrated Executable Architectures*, BookSurge Publishing, 2006.

[Date03] Date, C. J., *An Introduction to Database Systems*, 8th Edition, Addison Wesley, 2003.

[Denn08] Dennis, A. et al., *Systems Analysis and Design*, 4th Edition, Wiley, 2008.

[Dori95] Dori, D., "Object-Process Analysis: Maintaining the Balance between System Structure and Behavior," *Journal of Logic and Computation* 5(2), pp.227-249, 1995.

[Dori02] Dori, D., *Object-Process Methodology: A Holistic Systems Paradigm*, Springer Verlag, New York, 2002.

[Dori16] Dori, D., *Model-Based Systems Engineering with OPM and SysML*, Springer Verlag, New York, 2016.

[Enge02] Gregor Engels, G. Et al., "Consistency-preserving Model Evolution Through Transformations," *Proc. Int'l Conf. UML 2002*, pp. 212–227, 2002.

[Hoar85] Hoare, C. A. R., *Communicating Sequential Processes*, Prentice-Hall, 1985.

[Lale08] Laleau, R. et al., "Using Formal Metamodels to Check Consistency of Functional Views in Information Systems Specification," *Information & Software Technology*, pp. 797-814, 2008.

[Lin19] Lin, K. et al., "The Structure-Behavior Coalescence Approach for Systems Modeling," *IEEE Access*, Vol. 7, pp. 8609-8620, 2019.

[Malg06] Malgouyres, H. et al., "A UML Model Consistency Verification Approach Based on Meta-modeling Formalization", *Proceedings of the 2006 ACM*

Symposium on Applied Computing, pp. 1804-1809, 2006.

[Maie09] Maier, M. W., *The Art of Systems Architecting*, 3rd Edition, CRC Press, 2009.

[Miln89] Milner, R., *Communication and Concurrency*, Prentice-Hall, 1989.

[Miln99] Milner, R., *Communicating and Mobile Systems: the π-Calculus*, 1st Edition, Cambridge University Press, 1999.

[OMG 13a] OMG, Semantics of a Foundational Subset for Executable UML Models (fUML). *Object Management Group*, Needham, MA, 2013.

[OMG 13b] OMG, Action Language for Foundational UML (Alf). *Object Management Group*, Needham, MA, 2013.

[O'Rou03] O'Rourke, C. et al, *Enterprise Architecture Using the Zachman Framework*, 1st Edition, Course Technology, 2003.

[Pele00] Peleg, M. et al., "The Model Multiplicity Problem: Experimenting with Real-Time Specification Methods". *IEEE Tran. on Software Engineering*. 26 (8), pp. 742–759, 2000.

[Przi16] Przigoda, N. et al., "Analyzing Inconsistencies in UML/OCL Models", *Journal of Circuits, Systems and Computers*, 25(3), 2016.

[Rayn09] Raynard, B., *TOGAF The Open Group Architecture Framework 100 Success Secrets*, Emereo Pty Ltd, 2009.

[Roza11] Rozanski, N. et al., *Software Systems Architecture: Working With Stakeholders Using Viewpoints and Perspectives*, 2nd Edition, Addison-Wesley Professional, 2011.

[Rumb91] Rumbaugh, J. et al., *Object-Oriented Modeling and Design*, Prentice-Hall, 1991.

[Sang03] Sangiorgi, D. et al., *The Pi-Calculus: A Theory of Mobile Processes*, Cambridge University Press, 2003.

[Weil08] Weilkiens, T., Systems Engineering with SysML/UML: Modeling, Analysis, Design. Morgan Kaufmann, 2008.

[Weil08] Weilkiens, T., Systems Engineering with SysML/UML: Modeling, Analysis, Design. Morgan Kaufmann, 2008.

INDEX

A

abstract syntax, 17

action, 33, 34, 35, 36

  called action, 35

  calling action, 33, 34

action language for foundational UML, 20

activity diagram, 73

activity diagram relation, 74

actor, 32

AD. *See* activity diagram

ADR. *See* activity diagram relation

agent

  callee, 32

  caller, 32

Alf. *See* action language for foundational UML

B

building blocks. *See* object

C

callee, 32

caller, 32

channel-based infinite-queue SBC process algebra, 24

channel-based multi-queue SBC process algebra, 24

channel-based single-queue SBC process algebra, 24

C-I-SBC-PA. *See* channel-based infinite-queue SBC process algebra

class diagram, 91

class diagram relation, 91

ClsD. *See* class diagram

ClsDR. *See* class diagram relation

C-M-SBC-PA. *See* channel-based multi-queue SBC process algebra

ComD. *See* communication diagram

ComDR. *See* communication diagram relation

communication diagram, 85

communication diagram relation, 87

component. *See* object

concrete syntax, 17

conditional process, 26

C-S-SBC-PA. *See* channel-based single-queue SBC process algebra

E

entity. *See* object

external environment, 40

F

fix. *See* recursion

foundational UML, 20

fUML. *See* foundational UML

G

generalized SBC process algebra, 23

G-SBC-PA. *See* generalized SBC process algebra

I

IFD. *See* interaction flow diagram

inconsistency, 17, 18
interaction
  operation call interaction, 32
  operation return interaction, 32
interaction flow diagram, 45, 46

L

labelled transition system, 49
LTS. *See* labelled transition system

M

MDE. *See* model-driven engineering
metamodel, 17, 20
model
  metamodel, 17
  user model, 17
model-driven engineering, 17
model-driven software engineering. *See* model-driven engineering
multi-diagram, 17, 18
multiple diagrams, 18

N

non-aggregated system. *See* object
null process, 26

O

object, 40
object constraint language, 20
object diagram, 95
object diagram relation, 96
OCL. *See* object constraint language
OD. *See* object diagram
ODR. *See* object diagram relation
O-I-SBC-PA. *See* operation-based infinite-queue SBC process algebra
O-M-SBC-PA. *See* operation-based multi-queue SBC process algebra
operation, 31
operation call interaction, 32
operation return interaction, 32
operation signature, 32
operation-based infinite-queue SBC process algebra, 24
operation-based interaction
  operation call, 33
  operation return, 34
operation-based multi-queue SBC process algebra, 24
operation-based single-queue SBC process algebra, 24
orthogonal composite state, 58

P

parallel composition, 25
part. *See* object
port, 33, 34, 35
  called, 35
  calling, 33, 34
prefix, 50
process algebra
  channel-based infinite-queue SBC process algebra, 24
  channel-based multi-queue SBC process algebra, 24
  channel-based single-queue SBC process algebra, 24
  generalized SBC process algebra, 23
  operation-based infinite-queue SBC process algebra, 24
  operation-based multi-queue SBC process

algebra, 24

operation-based single-queue SBC process

algebra, 24

projection, 68, 72, 77, 84, 90, 93, 98

R

recursion, 26

replication, 26

S

SBC. *See* structure-behavior coalescence

sequence diagram, 79

sequence diagram relation, 81

sequentialization, 25

software behavior, 39, 40

software structure, 39, 40

SqD. *See* sequence diagram

SqDR. *See* sequence diagram relation

state

orthogonal composite state, 58

process expression, 58

state diagram, 69

state diagram relation, 70

StD. *See* state diagram

StDR. *See* state diagram relation

STOP. *See* null process

structure-behavior coalescence, 39, 40

summation, 25

syntax

abstract syntax, 17

concrete syntax, 17

systems model. *See* user model

T

TG. *See* transition graph

TGR. *See* transition graph relation

transition graph, 57

transition graph relation, 58

U

UC. *See* use case

UCD. *See* use case diagram

UCDR. *See* use case diagram relation

UML. *See* unified modeling language

unified modeling language, 17

use case, 65

use case diagram, 65

use case diagram relation, 66

user model, 17

www.ingramcontent.com/pod-product-compliance
Lightning Source LLC
Chambersburg PA
CBHW081000170526
45158CB00010B/2854

* 9 7 8 1 0 7 8 0 6 8 5 2 9 *